INTRACELLULAR CONSEQUENCES OF AMYLOID IN ALZHEIMER'S DISEASE

T0282345

INTRACELLULAR CONSEQUENCES OF AMYLOID IN ALZHEIMER'S DISEASE

MICHAEL R. D'ANDREA
President, Slidomics, LLC, Wilmington, DE, USA

AMSTERDAM • BOSTON • HEIDELBERG • LONDON
NEW YORK • OXFORD • PARIS • SAN DIEGO
SAN FRANCISCO • SINGAPORE • SYDNEY • TOKYO

Academic Press is an imprint of Elsevier

Academic Press is an imprint of Elsevier
125 London Wall, London EC2Y 5AS, UK
525 B Street, Suite 1800, San Diego, CA 92101-4495, USA
50 Hampshire Street, 5th Floor, Cambridge, MA 02139, USA
The Boulevard, Langford Lane, Kidlington, Oxford OX5 1GB, UK

ISBN: 978-0-12-804256-4

British Library Cataloguing-in-Publication Data
A catalogue record for this book is available from the British Library.

Library of Congress Cataloging-in-Publication Data
A catalog record for this book is available from the Library of Congress.

For Information on all Academic Press publications
visit our website at http://store.elsevier.com/

Working together
to grow libraries in
developing countries

www.elsevier.com • www.bookaid.org

Publisher: Mara Conner
Senior Acquisition Editor: Natalie Farra
Senior Editorial Project Manager: Kristi Anderson
Production Project Manager: Julia Haynes
Designer: Matt Limbert

Typeset by MPS Limited, Chennai, India

Dedication

I dedicate this book to:

- Those who passed away from this dreadful disease, and those who caringly donated their loved ones' tissues to research. None of us would have discovered so much without their ultimate contributions; and those who have Alzheimer's disease and their family's caregivers, for help is soon on the way.
- My wife Patty, my oldest daughter Dr Michelle and her husband, Kevin, my son Michael, and youngest daughter, Stephanie, all of whom have given me endless support and love.
- My parents, Henry and Angela, for all their eternal love and support, and especially for buying my first microscope as a Christmas gift when I was 7 years old that set my scientific career in motion.

Contents

About the Cover

The figures below represent serial sections of human Alzheimer's disease brain tissues processed for immunohistochemistry using an antibody to Aβ42 and using a negative control antibody. All nuclei are stained blue with hematoxylin. Arrowheads demonstrate the alignment of a set of five neurons in the serial sections.

In the left panel (and book cover), Aβ42-positive immunolabeling is presented as brown staining. Prominent Aβ-positive immunolabeling is present in the vascular smooth muscle cells (*large solid red arrow*), and in many cells, including neurons (*arrowheads*) and astrocytes (*small arrow*). An Aβ42-positive plaque is detected (*open arrow*). Also note the presence of several unstained red blood cells in the vessel (*stained pale blue*), as well as extracellular Aβ42-positive immunolabeling in a diffuse plaque that seems to leak from the Aβ42-laden vessel (at the 3:00 position of the vessel).

In the right panel, no detectable Aβ42 immunolabeling is observed in the negative control, including the area of the unstained amyloid plaque (*open arrow*). The asterisk near the neuron in the negative control shows the presence of intraneuronal lipofuscin (yellow-brown pigment that is typically observed in neurons as granules composed of lipid-containing residues of lysosomal digestion). (40x objective; courtesy of Slidomics, LLC.)

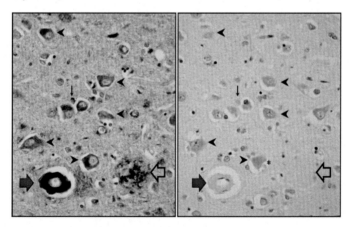

About the Author

Michael R. D'Andrea received his PhD in Cell and Developmental Biology and MS in Molecular Biology. He has authored over 100 scientific publications, including invited review papers on Alzheimer's disease (AD), and coinvented 11 patents. His technical expertise is in the areas of histopathology/neuropathology, immunohistochemistry, and image analysis. Since 1996, he was Team Leader and Principal Scientist of Target Validation Team at Johnson & Johnson Pharmaceutical Research & Development. There he discovered and validated novel targets, biomarkers, and compounds to treat cancer, inflammatory diseases, and AD, and accepted numerous awards for these endeavors. Currently, he is president and chief histopathologist at Slidomics, LLC.

He has presented his Alzheimer's research at the following sponsored international, national, and regional meetings: Society of Neuroscience; International Conference on AD and Related Disorders; The Alzheimer's Imaging Consortium; and International Neurodegeneration in AD, Parkinson's Disease & Related Disorders. In addition, he spoke at various meetings at the Annual Biological Staining Commission, The National Disease Research Institute, University of Pennsylvania, and was invited to lead the AlzForum's WebCast International discussion for the AD Forum on the evidence that neuronal cell death in AD is due to an autoimmune mechanism. He was also invited to the Challenging Views of AD: Round II meeting to debate the inflammatory aspects of AD. In addition, he has reviewed international AD grants (Spain, Israel) and is on several scientific editorial boards.

He was one of the first to relate the presence of intracellular $A\beta42$ in AD neurons to senile, dense-core plaque formation not by deposition, but through neuronal lysis that triggers the inflammatory responses that inflict to additional neuronal death, independent of $A\beta$. He characterized the existence of plaques types based on their origin of formation and provided histopathological evidence of apoptotic neuronal death through an autoimmune mechanism in AD, suggesting that AD is an autoimmune disease.

Most recently, he published a book entitled *Bursting Neurons and Fading Memories: An Alternative Hypothesis of the Neuropathology of AD*. Furthermore, Michael has animated the "Inside-Out" (or "Bursting") hypothesis (available on YouTube (https://www.youtube.com/watch?v=_NTaGjQow1c), now renamed the "Lytic Model" in this book. Currently, he continues to post discussions on the matter.

Preface

Analyzing hundreds of publications on a nuanced topic, selecting articles accurately and coherently to reflect the research, and then to tie these together into a meaningful story to allow you, the reader, to access this compilation has been a challenging task for me. A related difficulty was deciding on the scope of this anthology. I will not try to represent the multitude of aspects of a single medical condition, but will focus on a single, critical aspect of a specific area of related research. This is to draw clarity and attention to an otherwise neglected and confused topic, and to develop a novel hypothesis from an unbiased presentation of established, documented facts.

As you picked up this book, you will know the medical condition in question is Alzheimer's disease (AD), and the specific area of research is the intracellular consequences of amyloid. Researchers and scientists studying AD will immediately recognize the importance of amyloid senile plaques, one of the earliest discovered pathological attributes of the debilitating condition, and a focal point of many clinical trials investigating a cure. Most of what we know about the pathology of AD is focused around the amyloid protein, or specifically $A\beta$, the smaller reportedly toxic fragment that comprises the plaques. Those studies taught us how $A\beta$ is generated and processed into additional fragments, how neurons secrete it, how the toxic $A\beta$ aggregates outside the neuronal cells to form plaques, and how these extracellular plaques eventually cause neuronal death leading to AD.

The sequence of neuronal secretions of $A\beta$ leading to neuronal death is the basis of the "amyloid cascade hypothesis" and of a number of clinical efforts aimed to reduce or remove levels of $A\beta$ in AD patients. Unfortunately, these clinical trials failed to alter the course of the disease even though significant reductions of $A\beta$ levels and plaques were observed. So if it seems that $A\beta$ does not hold the key in curing AD after all, then why spend your time reading a book about amyloid?

The failing strategy may not be amyloid per se, but the specific targeting of the "extracellular" amyloid.

"Intracellular" $A\beta$ in the neurons and other cells has often been ignored because while we do know a great deal about the properties of $A\beta$ thanks to the collective efforts of many gifted scholars, many of their perspectives have narrowly considered the protein only in its assigned "extracellular" role as defined by the amyloid hypothesis. In other words, although intracellular amyloid was detected in neurons, its presence was merely considered a source of the extracellularly secreted amyloid.

Intracellular Aβ may have a more significant relationship to neuronal death and disease than the extracellularly deposited Aβ. This single and once unpopular change of perspective could be the key to all of the pathological events leading to AD and is worth the time to understand before turning our sights away from Aβ entirely.

To set the stage, chapters "Amyloid Basis of Alzheimer's Disease" and "Origin(s) of Intraneuronal Amyloid" present a brief historical background of amyloid, how Aβ is generated and processed, how it relates to the amyloid cascade hypothesis, and most importantly, the possible origins of the intraneuronal amyloid. Chapter "Natural Intracellular Consequences of Amyloid" analyzes the unfamiliar characteristics of Aβ as a necessary component in a myriad of natural physiological processes in neuronal and nonneuronal cells, while chapters "Pathological Consequences of Aβ from Extracellular to Intraneuronal," "Intraneuronal Amyloid and Plaque Formation," "Intraneuronal Amyloid and Cognitive Impairment," and "Intraneuronal Amyloid and Inflammation" present information of the emerging pathological aspects of *intraneuronal* amyloid and its relationship to amyloid plaque formation, to cognitive impairment, and inflammation, respectively. In chapter "Consequences of Intracellular Amyloid in Vascular System", the *intracellular* consequences of Aβ are presented in nonneuronal cells such as the smooth muscle cells and pericytes as related to pathologies in the AD vascular system. In closing, the implications of intracellular Aβ are discussed in chapter "Implications of Intraneuronal Aβ".

While the role of intracellular amyloid in Alzheimer's disease remains undiscovered, an increasing body of work suggests that the early pathogenesis of Alzheimer's begins with the accumulation of Aβ in neurons leading to their demise, and subsequent neuronal death independent of Aβ. However, upstream from neuronal death may be cell death of the vascular cells due to overaccumulation of Aβ as well, leading to a dysfunctional blood-brain barrier that leaks toxic levels of Aβ into the brain. Thus, I present this anthology to bolster research endeavors in this area to cure AD.

Acknowledgments

Especially to my immediate family for providing direction, style, and edits for the book.

To all of the hundreds of esteemed colleagues cited in this book and to those unintentionally overlooked; this book is a treasure of their contributions, which is a testament of their precious time dedicated to help cure Alzheimer's disease. And for those authors who committed many exhaustive hours publishing their review articles. This book would not have been written if it wasn't for their work.

Thank You!

Introduction

Alzheimer's disease (AD): ultimately waiting for most people if they live long enough. The statistics prove it, and chances are that you know someone personally affected by this dreadful disease. As of 2015, there are 5.3 million sufferers in the United States, and 47.5 million worldwide; that's more than the total population of Canada.[1] Just this single statistic alone is staggering. It is estimated that about 75.6 million people worldwide will have Alzheimer's in the next 15 years if there is no cure. The cause(s) of this progressive and fatal disease remain(s) unknown, and in spite of decades of research and bountiful funding, science has only been able to treat the symptoms without stopping the deleterious progression of the disease toward death.

AD is mostly associated with advanced age, and the clinical presentations of this neurodegenerative disorder are quite clear: marked decline in memory and cognitive performance, including impairment of language, visual and motor coordination, calculation, judgment and planning, and problem solving, leading to eventual death.[2,3] The neuropathological features are also quite clear: significant synaptic loss (20–50%), neuronal death throughout many areas in the brain, with particular degeneration of neurons in the cerebral cortex and hippocampus, intracellular neurofibrillary tangles, senile amyloid plaques, and gliosis.[4]

Just about every scientific paper, review, and editorial begins with the same text citing these AD hallmarks while noting "the cause(s) of this disease remain(s) unknown." Hence, hypotheses need to be generated to help determine the cause(s) of AD in an effort to provide a therapeutic opportunity for the cure.

Of the many hypotheses presented over the 20+ years, one hypothesis has generated the most interest and support, the amyloid cascade hypothesis. Hundreds of publications convincingly portrayed Aβ and Aβ42, the smaller fragmented amyloid species, as toxic, despite prior reports that seemed to leave little doubt that Aβ serves essential roles in synaptic plasticity, learning, memory, and survival, not to mention nonneuronal roles in cellular proliferation, wound healing, and more recently, as an antimicrobial.

The reports portraying the toxic nature of these amyloid precursor protein (APP) products are the building blocks of the amyloid hypothesis, which states that increased Aβ production and subsequent secretion by the neurons (or failure of Aβ clearance) induces a gradual Aβ accumulation in the brain through life, leading to the formation of extracellular amyloid senile plaques. These plaque formations induce inflammatory responses resulting in synaptic damage, neurofibrillary tangles, neuronal loss, then AD.[4] Therefore, based on the compelling body of scientific publications, it is logical to propose that inhibiting Aβ production should cure Alzheimer's, a paradigm that continues to drive several current clinical trials.[5]

One approach is to treat AD patients with active immunotherapy by stimulating the immune system to attack the Aβ in the brain.[6] However, this approach failed due to the development of serious side effects including the development of meningoencephalitis, transient vasogenic edema, and microhemorrhage.[5,7,8] Long-term follow-up of these patients showed a reduction of plasma Aβ levels and a clearance of amyloid plaques in the autopsied brain tissues. In spite of the successful target engagement, the degree of neurodegeneration was unchanged in these patients suggesting that lowering Aβ levels and removing amyloid plaques are not the targets to treat AD.[9]

Passive immunotherapy was also used to treat patients with AD by reducing extracellular Aβ.[1,5,6] Two of the Phase 3 clinical trials successfully lowered Aβ levels in plasma and cerebral spinal fluid (CSF) in patients with mild-to-moderate AD but without significant improvement in primary outcome measures, such as improved cognition or functional ability.[2] Despite successful target engagement once again, the inability to demonstrate clinical benefit suggests that reducing extracellular amyloid may not be the correct target to cure AD.[2,7]

A third approach in treatment was targeting the enzymes along the amyloid processing pathway (eg, γ-secretase inhibition and γ-secretase modulation) to lower or inhibit the Aβ production, but this therapy also failed to show benefit to the patients, further suggesting that reducing its production and/or level may not be the key to curing AD as was once so commonly thought.[10]

Is "extracellular" Aβ the correct target to cure AD; is the amyloid cascade hypothesis "too big to fail"?[11]

The clinical data show that removing "extracellular" Aβ or reducing its production will not cure AD. However, more and more recent research suggests that the way to save the neurons is to inhibit the intracellular, not extracellular, toxic accumulations of Aβ. This is the essence of this book: to first present studies of the essential roles of intracellular Aβ in the many biological processes of life, and then to present studies concerning the relationship of intraneuronal Aβ to neuronal death, to plaque formation, to cognition, and to inflammation in the brain with additional attention to the roles of intracellular Aβ in nonneuronal cells.

References

1. <http://www.alzheimer.ca/en/About-dementia/What-is-dementia/Dementia-numbers>. (downloaded July 7, 2015).
2. Cummings JL. Review. Alzheimer's disease. *N Engl J Med* 2004;**351**:56—67.
3. McKhann G, Drachman D, Folstein M, Katzman R, Price DL, Stadlan EM. Review. Clinical diagnosis of Alzheimer's disease: report of the NINCDS-ADRDA Work Group under the auspices of Department of Health and Human Services Task Force on Alzheimer's disease. *Neurology* 1984;**34**:939—44.
4. Xiaqin S, Yan Z. Amyloid Hypothesis and Alzheimer's Disease. In: Chang R C-C, editor. *Advanced Understanding of Neurodegenerative Diseases*. InTech; 2011. ISBN: 978-953-307-529-7, Available from: <http://www.intechopen.com/books/advanced-understanding-of-neurodegenerativediseases/amyloid-hypothesis-and-alzheimer-s-disease>.
5. Lambracht-Washington D, Rosenberg RN. Review. Advances in the development of vaccines for Alzheimer's disease. Discovery. *Medicine* 2013;**15**(84):319—26.
6. Lobello K, Ryan JM, Liu E, Rippin G, Black R. Review. Targeting β amyloid: a clinical review of immunotherapeutic approaches in Alzheimer's disease. *Intern J Alzheimers dis* 2012;**2012**:628070.
7. Lemere CA. Review. Immunotherapy of Alzheimer's disease: hoops and hurdles. *Mol Neurodegen* 2013;**8**:36.
8. Pfeifer M, Boncristiano S, Bondolfi L, Stalder A, Deller T, Staufenbiel M, et al. Cerebral hemorrhage after passive anti-Aβ immunotherapy. *Science* 2002;**298**:1379.
9. Holmes C, Boche D, Wilkinson D, Yadegarfar G, Hopkins V, Bayer A, et al. Long-term effects of Aβ42 immunisation in Alzheimer's disease: follow-up of a randomised, placebo-controlled phase I trial. *Lancet* 2008;**372**:216—23.
10. Butterfield DA, Boyd-Kimball D, Castegna A. Review. Proteomics in Alzheimer's disease: insights into potential mechanisms of neurodegeneration. *J Neurochem* 2003;**86**:1313—27.
11. Castellani RJ, Smith MA. Review. Compounding artefacts with uncertainty, and an amyloid cascade hypothesis that is "too big to fail." *J Pathol* 2011;**224**(2):147—52.

CHAPTER

1

Amyloid Basis of Alzheimer's Disease

OUTLINE

HISTORICAL BACKGROUND OF AMYLOID AND PLAQUES

Amyloid is a somewhat infamous insoluble, fibrous protein; infamous because amyloid deposits are responsible for tissue damage in a fair number of genetic and inflammatory diseases and disorders.[1] Of course in modern times, amyloid has become most commonly associated with Alzheimer's disease (AD): as of November 2015, a Google search for "amyloid and Alzheimer's" gets almost 11 million hits.

Before the relationship between amyloid and Alzheimer's was discovered, there were accounts of the plaques in the brain that became associated with dementia. Early descriptions of nerve cell degeneration

Intracellular Consequences of Amyloid in Alzheimer's Disease.
DOI: http://dx.doi.org/10.1016/B978-0-12-804256-4.00001-2

in senile dementia were described as nodules of glial sclerosis, or round heaps of nerve cell degeneration.[2] Subsequently in 1898, Emil Redlich named miliary sclerosis as "plaques" in two cases of senile dementia.[3] He also described plaques of different sizes and forms like the smaller cotton-wool type suggesting they represented a modified glial cell.

Several years later in 1907, extracellular deposits (also known as senile or neuritic plaques) were described as miliary foci of dystrophic neuronal processes surrounding a "special substance in the cortex" by Alois Alzheimer in the autopsied brain of a 51-year-old patient, who presented a very unusual clinical picture with loss of short-term memory and odd behavioral symptoms.[4–7] However, it would not be until the application of the Congo red stain that this "special substance" would be identified as amyloid creating an association between amyloid, dementia, and senile plaques.[4,5] Using a silver staining method, Alzheimer also identified the presence of neurofibrillary deposits in sections of her autopsied brain tissues; these were subsequently determined to be composed of aggregates of the abnormally hyperphosphorylated tau protein.[4,8]

At the same time in 1907, Oskar Fischer provided the first illustrations of the neuritic plaques that captured many of the features reported today (Fig. 1.1).[9] Fischer studied a total of 275 brains from cases of psychosis, neurosyphilis, and controls of various ages, with 110 being over 50 years old at the time of death. He observed plaques in 56 cases, all of whom

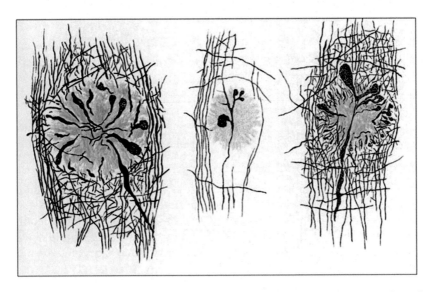

FIGURE 1.1 Drawings of three neuritic plaques from the brains of patients with senile dementia. Compiled from the illustrations of Fischer's 1907 paper. Note the abnormal, club-shaped neurites and the displacement of normal-looking fibrils in the space occupied by the plaques. *Source: Used with permission from Brain 2009;132:1102—11.*

were >50 years of age.[3] Alzheimer and Fischer disagreed on the origin of tangles: Alzheimer believed tangles consisted of chemically modified neurofibrils, while Fischer thought that they represented fibril proliferation *de novo*, and that the material was unrelated to neurofibrils.[3] However, the origin of the plaque remains a matter of debate (see Chapter 5).

Despite these discoveries of plaques in the demented brain, it took additional effort to recognize amyloid as the component of these deposits. Amyloid itself was first discovered over 150 years ago through an iodine-sulfuric acid test that demonstrated the transformation of plant material to starch.[4,10] This starch-like material was for the first time, referred to as amyloid, from the Latin word for starch, "amylym."[11]

Even though such amyloid deposits may have been observed as homogenous material in the liver and spleen back in 1693, it wasn't until 1854 that amyloid was first described as small round deposits in the nervous system using the same iodine-sulfuric acid staining method.[12] Although the iodine-sulfuric acid test produced similar results to the newly developed metachromatic stain in 1878, eventually the method of choice to detect amyloid in tissue sections would become the Congo red stain, initially produced to stain textile fibers.[13,14] The application of the Congo red staining led to the discovery of cerebrovascular amyloid in 80–90% in the brains of patients with AD.[15] Over time, the standard staining methods to describe AD pathology would be the silver stain to identify tangles, along with the Congo red to stain amyloid in microscopic sections of autopsied brain tissues.[4]

Although several biochemical methods led to the discovery of several forms of amyloidosis, it was a new water extraction method that led to the discovery of different amyloid proteins.[4] After years of describing various forms of amyloid, it was determined that the plaques in AD brains were composed of insoluble protein denaturants, and did not represent any known form of amyloid at that time.[16] Then in 1984, the AD-associated Aβ form of amyloid (Aβ), 4.2-kDa peptide, primarily 40 or 42 amino acids in length, in the cerebrovascular tissue in Alzheimer's tissues was also found in a Down syndrome patient, thereby providing a link between the two conditions.[4,17,18] The next year, the same Aβ was described in the AD plaques, and later determined to be toxic (see Chapter 4).[4,19,20]

Amyloid Generation and Processing

Of all the neuropathological features reported in the AD brain, such as neuronal and synaptic loss, and NFTs, it is the extracellular aggregates of Aβ peptides as senile plaques that have occupied most of the AD research activity, especially concerning how Aβ is generated and processed from APP. Such research has been in an effort to prevent aggregation, and spare neuronal death.

APP is a single-pass transmembrane protein, referred to as a type I integral membrane protein (Fig. 1.2), and is encoded by a single gene on human chromosome 21, containing 18 exons.[7,21,22] APP has a signal peptide, a large extracellular N-terminal domain, a small intracellular

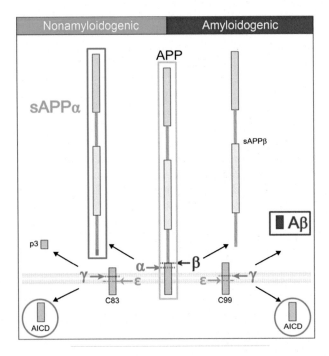

FIGURE 1.2   Cleavage of APP and physiological roles of APP and APP fragments. APP can be cleaved via two mutually exclusive pathways. In the so-called amyloidogenic pathway APP is cleaved by BACE1 and γ-secretase enzymes (presenilin I is the catalytic core of the multiprotein γ-secretase complex). The initial β-secretase cleavage produces a large soluble extracellular domain, sAPPβ. The remaining membrane bound C99 stud is then cleaved by multiple sequential γ-secretase cleavages. These begin near the inner membrane at a γ-secretase cleavage site ε (the ε-site) to produce the AICD, and then subsequent sequential γ-secretase cleavages trim the remaining membrane bound component to produce different length Aβ peptides including Aβ43, Aβ42, Aβ40, and Aβ3. In the so-called nonamyloidogenic pathway APP is processed consecutively by α- and γ-secretases to produce sAPPα, p3 (which is in effect Aβ17-40/42), and AICD. The major α-secretase enzyme is ADAM10. Cleavage via amyloidogenic and nonamyloidogenic pathways depends on the cellular localization of cleavage enzymes, and of full-length APP, which are expressed and trafficked in specific subcellular locations. *APP*, Amyloid Precursor Protein; *AICD*, APP intracellular domain; *BACE1*, β-site APP cleaving enzyme; *sAPPβ*, secreted Amyloid Precursor Protein-β; *ADAM10*, A Disintegrin and Metalloproteinase domain-containing protein 10. *Source: Has been slightly modified: Used with permission from* Acta Neuropathol Commun 2014; 2:135.

C-terminal domain, a single transmembrane domain, and an endocytosis signal at the C-terminal.[7] APP not only has a very wide distribution in the body, but is expressed at high levels in the brain, is produced in large quantities in neurons, and is rapidly metabolized (see Chapter 3).[5]

APP undergoes enzymatic processing to produce fragments some of which are believed to play a crucial role in the pathogenesis of AD. This is not only because some of these fragments are located in the senile plaques, but also because mutations in the APP gene and the processing enzymes (eg, presenilin I, the catalytic core of γ-secretase complex) have been associated with rare cases of familial and inherited early onset of AD, respectively.[6,15,21,23–26] APP is cleaved sequentially by α-, β-, and γ-secretases, which results in the generation of the large soluble NH_2-terminal ectodomain, small hydrophobic extracellular Aβ (Aβ40- and Aβ42-residues) peptide, and APP intracellular domain (AICD, 57- and 59-residue-long COOH-terminal fragments).[5,19–21,27]

The cleavage and processing of APP is divided into the nonamyloidogenic and amyloidogenic pathways (Fig. 1.2).[6,7,20,25] In the nonamyloidogenic pathway, APP is first cleaved by the α-secretase producing the soluble APP-α (sAPP-α) peptide that is secreted into the extracellular medium.[28] The intact membrane fragment is subsequently cleaved by γ-secretase at two areas in the remaining fragment to produce a short fragment (p3) and the AICD. Hence, this nonamyloidogenic pathway does not produce the Aβ peptide.

The Aβ peptide is produced through the processes of the amyloidogenic pathway (Fig. 1.2). APP is first cleaved by β-secretase yielding two products: the soluble APP-β fragment and the membrane-retained fragment. This membrane product of APP is subsequently cleaved by γ-secretase to produce the Aβ peptide and another AICD fragment. Mutations in these processes that are associated with AD are discussed in Chapter 4.

Fig. 1.3 shows how APP traffics in the neurons, and how APP is associated with several cellular organelles. After sorting in the endoplasmic reticulum and Golgi, APP is delivered to the axon and further processed in the membrane.[5,26]

Aβ States

Once released from the cell, the Aβ peptides contain an amino acid sequence that favors Aβ–Aβ binding and β-sheet formation.[29] Aβ aggregates with other Aβ peptides forming a variety of shapes and molecular weights (Fig. 1.4) that may impact its toxicity.[27,30] The formation and biological activities of Aβ oligomers, protofibrils, and fibrils have been under intensive investigation in recent years. Although insoluble fibrillar Aβ has been shown to be neurotoxic, compelling evidence also indicates that

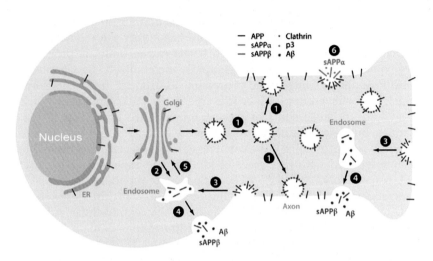

FIGURE 1.3   APP trafficking in neurons. Newly synthesized APP (*purple*) is transported from the Golgi down the axon (1) or into a cell body endosomal compartment (2). After insertion into the cell surface, some APP is cleaved by α-secretase (6) generating the sAPPα fragment, which diffuses away (*green*), and some is reinternalized into endosomes (3), where Aβ is generated (*blue*). Following proteolysis, the endosome recycles to the cell surface (4), releasing Aβ (*blue*) and sAPPβ. Transport from the endosomes to the Golgi prior to APP cleavage can also occur, mediated by retromers (5). *APP*, Amyloid Precursor Protein; *sAPPα*, soluble APPα. *Source: Used with permission from* Annu Rev Immunol 2011;**34**:185−204.

oligomers and protofibrils contribute significantly to cellular cytotoxicity, inflammatory responses, synaptic dysfunction, and reduced neurogenesis.[29] Hence, the aggregation of the Aβ species is thought to play a pivotal role in the disease progression of AD through a cascade of events, as described in the amyloid cascade hypothesis (see Chapter 4).[8,26]

The kinetic relationship is not clear among the two Aβ species: Aβ40 and Aβ42.[30] Although oligomers may be the intermediate in fibril formation (Fig. 1.4), it is possible that oligomers may actually represent a separate assembly pathway.[31] For example, the aggregation conditions for Aβ42 did not compare to structural or functional species of Aβ40, even though the Aβ42 peptide is more fibrillogenic than the Aβ40 species. The differences may be attributed to aggregation time, assuming they follow the same pathways.[30,32] The pH also affects the aggregation conditions of the Aβ peptides. The metal binding sites of Aβ contain three histidine residues of Aβ that are involved in the interaction with metal ions, and the metal-His(Ntau) ligation is a common feature among the insoluble Zn(II)- and Cu(II)-Aβ aggregates at pH 5.8-7.4 and 5.8-6.6, respectively.[33] Interestingly, under normal physiological conditions, Cu(II) is expected to protect Aβ against Zn(II)-induced aggregation by competing with Zn(II) for histidine residues of Aβ.

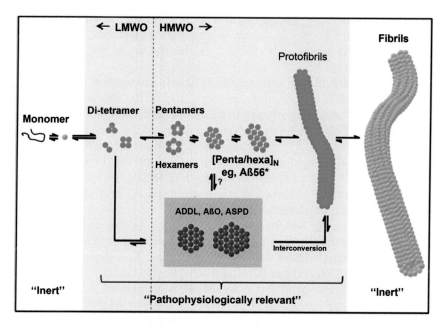

FIGURE 1.4 Pathways of aggregation and observed Aβ-aggregate intermediates. Monomeric Aβ folds to the activated state and then exists in rapid equilibrium with LMWO, which aggregate over various transient high molecular weight intermediates to matured fibrils. The definition of LMWO and HMWO is related to the elution profile of Aβ-aggregates in size exclusion chromatography, revealing two predominant peaks at the exclusion limit (>60 kDa) and at the void volume (4–20 kDa), respectively. The HMW intermediates comprise pentamers, hexamers, and multiples thereof, finally forming protofibrils, which are the precursors for multistranded ribbons of matured fibrils. Further neurotoxic aggregate species for example AβO, ADDL, and ASPD are believed to aggregate over alternative pathways but preliminary data revealed that these are able to converge into the other pathways of aggregation (interconversion). Interestingly, every change in the experimental paradigm can provoke this aggregate conversion. Therefore, one might assume that many different aggregates coexist and, thus, neurotoxicity can be attributed to several pathogenic modes of action. Monomers and fibrils are believed to be biologically inert; however fibrils are able to collapse into protofibrils and then also reveal toxicity. The broad range of prefibrillar aggregates have been reported as pathophysiologically relevant in AD. *AβO*, Amyloid-β-oligomers; *ADDL*, Alzheimer-derived diffusible ligands; *ASPD*, amylospheroids; *HMWO*, high molecular weight oligomers; *LMWO*, low molecular weight oligomers. *Source: Used with permission from* Immun Ageing 2013;*10:18.*

DETECTING INTRANEURONAL AMYLOID

APP was reported in neurons over 30 years ago as the source of the extracellular amyloid in the AD brains, whereby the neurons secrete the Aβ to eventually form toxic senile plaques. These eventually lead to the demise of the neurons, as per the amyloid cascade hypothesis (see Chapter 4). Given these theories of events, the presence of Aβ in neurons was only viewed as the source of the extracellular amyloid.

But in the last 15 years, that attitude slowly changed as the attention turned back inside the neurons, perhaps due to the inaccuracies of the amyloid hypothesis. Reports began to demonstrate how Aβ accumulates in neurons, and how that may be considered one of the earlier pathological events leading to AD.[26,34–42] Subsequently, these early observations led to hundreds of follow-up papers suggesting that "further investigations of intraneuronal Aβ could improve the understanding of early stage AD and the mechanistic links between intraneuronal Aβ and tau pathology, neurodegeneration and dementia."[43]

Before discussing the findings of these reports (the purpose of this book), it is important to review some of the technical details on how the data was generated. These data are typically based on the immunohisto-chemical (IHC) detection of Aβ in the cells. For the most part, the development of IHC has provided a wealth of contributions to help propel discoveries of the pathological processes leading to AD, and it was my expertise in this methodology that facilitated my contributions in this field.[44] Essentially, IHC is a method of staining specific targets in tissues for microscopic analyses. Unlike typical slide-staining methods that stain tissue elements indiscriminately (eg, haematoxylin and eosin stain), IHC utilizes target-specific antibodies to visualize their specific antigens in tissues.

The integrity of the method is dependent upon the specificity of the antibody to its antigen; otherwise, nonspecific binding can mislead the data, and therefore the interpretation of the results. In the case of using IHC to stain targets (eg, Aβ) in AD tissues, other variables can also contribute to erroneous staining, such as the degree and type of fixation of the tissue, the use of antigen retrieval pretreatment methods, and quality of the detection reagents to produce the staining. Indeed, the proper uses of positive and negative controls are essential, but in addition, the antibody has to be validated to its antigen. Although Western blot data supports the molecular weight characteristics, preincubating the antibody with its specific antigen is another method used to validate the specificity of the antibody. Hence, if the antibody is preincubated with its specific antigen, and then placed on the tissue sections mounted on microscopic slides for detection, no immunolabeling should be observed since the antibody-specific antigen binding site on the antibody should be occupied with the antigen, leaving no free binding sites on the antibody to bind to the antigens in the tissue.

APP/Aβ-Related Antibodies

Antibodies are specific to epitopes of the target, or antigens (*antibody generators*). Hence, any data discussing the presence of APP and Aβ in tissues will depend on the antibody selected.

As an example, seven different commercial antibodies (4G8, 6E10, 82E1, 6F3D, Aβ40, Aβ42, 12F4) directed to the N- or C-terminus or mid-portion of the Aβ fragment (Fig. 1.5) were used to demonstrate that the selection of the primary antibody is critical to the interpretation of the study.[45] All of these antibodies claim to recognize Aβ in the tissue according to their specifications. With three (4G8, 6E10, 82E1) of these seven antibodies, intracellular immunolabeling was detected in a

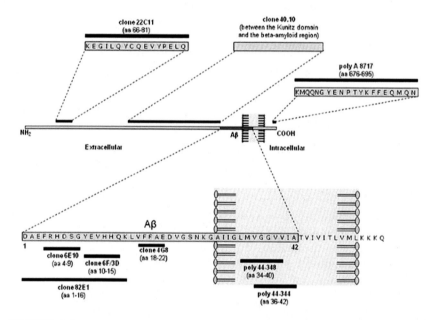

FIGURE 1.5 A schematic presentation of AβPP where the epitope regions recognized by the antibodies used in this study are marked with black (AβPP/Aβ) and gray bars (Aβ neoepitopes). Clone 82E1 is raised against $A\beta_{1-16}$ (Immuno-Biological Laboratories). Clone 6E10 is raised against $A\beta_{1-17}$ (Signet). Clone 6F3D is raised against $A\beta_{8-17}$ (DakoCytomation). Clone 4G8 (Signet) is raised against $A\beta_{17-24}$ (Signet). Clone 12F4 is raised against $A\beta_{1-42}$ (Covance) and is reactive to C-terminus of Aβ and is specific for the isoform ending at the 42nd amino acid. Clone 22C11, is raised against recombinant Alzheimer precursor A4 fusion protein; clone 40.10, is raised against the sequence between Kunitz protease inhibitor domain and the β-amyloid region (Novocastra). The specificity of polyclonal antibodies as provided by the manufacturer: poly 44−348 (Biosource/Invitrogen), poly 44−344 (Biosource/Invitrogen), and poly A 8717 (Sigma). *AβPP*, Amyloid-β Protein Precursor; *Aβ*, Amyloid-β. *Source: Used with permission from* J Alzheimers Dis 2010;20:1015−28.

variety of normal human tissues, including that of a 2-year-old, and AD brain tissues; similar immunolabeling was observed in transgenic mice brain tissues using the N-terminus antibodies (4G8, 6E10, 3D6).[43] However, only intracellular Aβ was detected in the AD tissues using the C-terminal antibodies (Aβ40, Aβ42, 12F4), while the monoclonal antibody 6F3D did not label the intracellular compartment with any of the tissues. Hence, for some of the data, the detection of intracellular Aβ in all tissues using the N-terminus antibodies represented a product of normal neuronal metabolism.[19,46,47] However, if only the antibodies directed to C-terminus of the Aβ had been used, the conclusion would have been that the intracellular Aβ is only detected in the brains of subjects with AD, and this would confirm the reports suggesting that the accumulation of intracellular Aβ is an event associated with the pathogenesis of AD.[45] These results strongly emphasize that staining results and interpretations are strongly dependent on several variables that include the choice of antibodies as well as the methods employed (eg, pretreatment condition) when assessing the presence of intracellular Aβ.

Fixation and Pretreatment Factors

Tissue fixation is used to chemically preserve the natural state of the tissue or cells for subsequent histological analyses. Unfortunately, the type (eg, formalin, paraformaldehyde) and duration of fixation can hamper the ability of the antibody to bind to its target antigen, which could lead to false-negative immunolabeling. For example, some studies have only found intracellular Aβ in the brains of aged and AD individuals, but not in the brains of nonhuman primates, while others have detected intracellular Aβ in cortical neurons of monkeys of various ages.[48] The author noted that the inconsistent data could be attributed to differences in tissue fixation, as the anti-Aβ antibodies produced a stronger immunolabeling signal when tissues were fixed with paraformaldehyde than formalin, which may be due to the cross-linkages in paraformaldehyde-fixed tissues that are much weaker than those in tissues fixed with neutral-buffered formalin.[49]

To combat the issues of fixation, several methods were designed to pretreat the specimens before the IHC assay.

Pretreatment methods on the tissue slides include the use of formic acid, periodic acid, enzymatic digestion, or an antigen retrieving/restoration method using heat to assist antibody penetration into the targets in the tissues or cells. In one example, a study compared three conditions (heat or enzymatic pretreatment to no pretreatment) on their effect to detect Aβ42 in formalin-fixed, paraffin-embedded human AD tissues.[50,51] Although all three protocols produced Aβ42 immunolabeling in amyloid plaques using four commercially obtained Aβ42 specific antibodies, only the heat pretreatment protocol consistently detected prominent intracellular Aβ42 in pyramidal neurons suggesting that consistent detection of intracellular Aβ42 is dependent on the IHC protocol using controlled heat.

The use of the formic acid pretreatment method was also evaluated on its effect to detect intracellular Aβ42 in control and AD brain tissues.[44,45,51] Both methods detected amyloid in plaques, neurons, ependymal cells, circulating monocytes, vascular smooth muscle, and endothelial cells. Although there were no observable differences in the intensity of the amyloid labeling in these cell types using both pretreatment methods, there were considerable differences in the intensity of amyloid immunolabeling in the plaques. The formic acid produced much more intense amyloid labeling in the plaques than the heat method perhaps overshadowing the relatively less intense and less relevant intraneuronal Aβ immunolabeling. With the heat method, the intensity of the amyloid labeling in the plaques was similar to that detected in nearby neurons, suggesting a neuronal origin of plaques. These data suggest that the obvious benefits of formic acid for increasing the intensity of amyloid plaque immunolabeling may unintentionally emphasize plaques over amyloid-containing cells during analyses especially considering that plaque load was typically the objective of the stain.[51]

In another study, when formic acid was used in conjunction with heat pretreatment, the formic acid treatment counteracted such effects of heat pretreatment via autoclaving. Thus, intraneuronal Aβ42 accumulation may have been underestimated by conventional methods using formic acid only.[52] However, contrary findings were reported noting that formic acid was preferred for the staining of highly aggregated Aβ peptides in fixed frozen or paraffin tissues of the AD transgenic mouse brain.[43,53]

The use of formic acid and/or heat pretreatment can also affect the intensity of immunolabeling by specific antibodies. Intracellular immunolabeling with clones 6E10 and 82E1 (Fig. 1.5) was only seen when the sections were both heated and incubated in formic acid.[45,54] In another study, heat pretreatment alone increased immunolabeling the 4G8 and AβPP antibodies, and so the need to properly annotate each step in the methods is so vital not only for reproducibility, but to help in the interpretation of the data.[45]

Critical technical factors, such as the type of tissue fixation, selection of the primary antibody, the type of pretreatment method (if any), and the detections system (although not discussed) will affect immunolabeling intensity, and not only affect the interpretation and reproducibility, but can make it challenging to compare data among studies, a bane that continues to impact IHC methods.

SUMMARY

Before delving into the biological properties of amyloid, the focus of this chapter is to provide a brief background of the origin of amyloid and plaques as they relate to AD. Not only is it important to understand the historical association of amyloid and AD, but it is also informative to understand how amyloid is produced, processed, and detected in cells, which is the basis of the data presented in this book.

References

1. Kumar V, Abbas AK, Fausto N, Mitchell RN. *Robbins Basic Pathology*. 8th ed. Philadelphia, PA: Saunders; 2007. p. 166.
2. Beljahow S. Pathological modifications in the brains of demented patients. *J Ment Sci* 1889;**35**:261−2.
3. Goedert M. Review. Oskar Fisher and the study of dementia. *Brain* 2009;**132**:1102−11.
4. Maarit T. "Amyloid" — Historical Aspects. In: Feng D, editor. *Amyloidosis*. InTech; 2013. ISBN: 978-953-51-1100-9. < http://dx.doi.org/10.5772/53423 >. Available from: < http://www.intechopen.com/books/amyloidosis/-amyloid-historical-aspects >.
5. O'Brien RJ, Wong PC. Review. Amyloid precursor protein processing and Alzheimer's disease. *Annu Rev Neurosci* 2011;**34**:185−204.
6. LaFerla FM, Green KN, Oddo S. Review. Intracellular amyloid-β in Alzheimer's disease. *Nature Rev Neurosci* 2007;**8**:499−509.
7. Xiaqin S, Yan Z. Amyloid Hypothesis and Alzheimer's Disease. In: Chang R C-C, editor. *Advanced Understanding of Neurodegenerative Diseases*. InTech; 2011. ISBN: 978-953-307-529-7, Available from: < http://www.intechopen.com/books/advanced-understanding-of-neurodegenerativediseases/amyloid-hypothesis-and-alzheimer-s-disease>.
8. Mohamed A, de Chaves EP. Review. Aβ internalization by neurons and glia. *Inter J Alz Dis* 2011;**127984**:17.
9. Fischer O. Miliare mekrosen mit drusigen wucherungen der neurofibrillen, eine regelmassige veranderung der hirnrinde bei seniler demenz. *Monatsschr Psychiat Neurol* 1907;**22**:361−72.
10. Aterman K. A historical note on the iodine-sulphuric acid reaction of amyloid. *Histochemistry*. 76;**49**(2):131−43.
11. Kyle RA. Review. Amyloidosis: a convoluted story. *Br J Haematol* 2011;**114**(3):529−38.
12. Virchow R. Über eine in gehirn und rückenmark des menschen aufgefundene substanz mit der chemischen reaction der cellulose. *Virchow's Archiv fur pathologische Anatomie und für klinische Medicin*, Berlin 1854;**6**:135−8.
13. Sipe JD, Cohen AS. Review: history of the amyloid fibril. *J Struct Biol* 2000;**130**(2−3):88−98.
14. Frid P, Anisimov SV, Popovic N. Congo red and protein aggregation in neurodegenerative diseases. *Brain Res Rev* 2007;**53**(1):135−60.
15. Pantelakis S. A particular type of senile angiopathy of the central nervous system: congophilic angiopathy, topography and frequency. *Monatsschr Psychiatr Neurol* 1954;**128**(4):219−56.
16. Allsop D, Landon M, Kidd M. The isolation and amino acid composition of senile plaque core protein. *Brain Res* 1983;**259**(2):348−52.
17. Glenner GG, Wong CW. Alzheimer's disease and Down's syndrome: sharing of a unique cerebrovascular amyloid fibril protein. *Biochem Biophys Res Commun* 1984;**122**(3):1131−5.

18. Glenner GG, Wong CW. Alzheimer's disease: initial report of the purification and characterization of a novel cerebrovascular amyloid protein. *Biochem Biophys Res Commun* 1984;**120**(3):885–90.
19. Parihar MS, Brewer GJ. Review. Amyloid beta as a modulator of synaptic plasticity. *J Alzheimers Dis* 2010;**22**(3):741–63.
20. Morris GP, Clark IA, Vissel B. Review. Inconsistencies and controversies surrounding the amyloid hypothesis of Alzheimer's disease. *Acta Neuropathologia Commun* 2014;**2**:15.
21. Russo C, Venezia V, Repetto E, Nazzari M, Violani E, Carlo P, et al. Review. The amyloid precursor protein and its network of interacting proteins: physiological and pathological implications. *Brain Res Rev* 2005;**48**:257–64.
22. Hardy J, Selkoe DJ. Review. The amyloid hypothesis of Alzheimer's disease: progress and problems on the road to therapeutics. *Science* 2002;**297**:353–6.
23. Christensen DZ, Schneuder-Axmann T, Lucassen PJ, Bayer TA, Wirths O. Accumulation of intraneuronal Aβ correlated with ApoE4 genotype. *Acta Neuropathol* 2010;**119**:555–66.
24. Rozemiller AJ, van Gool WA, Eikelenboom P. The neuroinflammatory response in plaques and amyloid angiopathy in Alzheimer's disease: therapeutic implications. *Curr Drug Targets CNS Neurol Disord* 2005;**4**(3):223–33.
25. Wang H, Megill A, He K, Kirkwood A, Lee H-K. Consequences of inhibiting amyloid precursor protein processing enzymes on synaptic function and plasticity. *Neural Plasticity* 2012;**272374**:24.
26. Echeverria V, Cuello AC. Review. Intracellular Aβ amyloid, a sign for worse things to come? *Mol Neurobio* 2002;**26**(2–3):299–316.
27. Moreth J, Mavoungou C, Schindowski K. Review. Passive anti-amyloid immunotherapy in Alzheimer's disease: what are the most promising targets?. *Immun Ageing* 2013;**10**:18.
28. Kojro E, Fahrenholz F. The non-amyloidogenic pathway: structure and function of α-secretases. *Subcell Biochem* 2005;**38**:105–27.
29. Lue L-F, Yan SD, Stern DM, Walker DG. Review. Preventing activation of receptor for advanced glycation endproducts in Alzheimer's Disease. *Curr Drug Targets CNS Neurol Disord* 2005;**4**:249–66.
30. Dahlgren KN, Manelli AM, Stine WB, Baker LK, Krafft GA, LaDu MJ. Oligomeric and fibrillar species of amyloid-β peptides differentially affect neuronal viability. *J Biol Chem* 2002;**35**:32046–53.
31. Kirkitadze MD, Condron MM, Teplow DB. Identification and characterization of key kinetic intermediates in amyloid β-protein fibrillogenesis. *J Mol Biol* 2001;**312**:1103–19.
32. LeVine H. Alzheimer's β-peptide oligomer formation at physiologic concentrations. *Anal Biochem* 2004;**335**:81–90.
33. Miura T, Suzuki K, Kohata N, Takeuchi H. Metal binding modes of Alzheimer's amyloid β-peptide in insoluble aggregates and soluble complexes. *Biochemistry* 2000;**39**:7024–34.
34. Masters CL, Multhaup G, Simms G, Pottgiesser J, Martins RN, Beyreuther K. Neuronal origin of a cerebral amyloid: neurofibriliary tangles of Alzheimer's disease contain the same protein as the amyloid of plaque cores and blood vessels. *EMBO J* 1985;**4**(11):2757–63.
35. Wirths O, Multhaup G, Czech C, Feldmann N, Blanchard V, Tremp G, et al. Intraneuronal APP/Aβ trafficking and plaque formation in β-amyloid precursor protein and presenilin-1 transgenic mice. *Brain Pathol* 2002;**12**:275–86.
36. Wirths O, Bayer TA. Review. Intraneuronal Aβ accumulation and neurodegeneration: lessons from transgenic models. *Life Sci* 2012;**91**:1148–52.
37. Hashimoto M, Bogdanovic N, Volkmann I, Aoki M, Winblad B, Tjernberg LO. Analysis of microdissected human neurons by a sensitive ELISA reveals a correlation

between elevated intracellular concentrations of Aβ42 and Alzheimer's disease neuropathology. *Acta Neuropathol* 2010;**119**:543−54.

38. Cheung C, Goh YT, Wu C, Guccione E. Modeling cerebrovascular pathophysiology in amyloid-β metabolism using neural-crest-derived smooth muscle cells. *Cell Rep* 2014;**9**:391−401.

39. Wirths O, Multhaup G, Bayer TA. Review. A modified β-amyloid hypothesis: intraneuronal accumulation of the β-amyloid peptide—the first step of a fatal cascade. *J Neurochem* 2004;**91**:513−20.

40. Nicoll JA, Yamada M, Frackowiak J, Mazur-Kolecka B, Weller RO. Cerebral amyloid angiopathy plays a direct role in the pathogenesis of Alzheimer's disease. ProCAA position statement. *Neurobiol Aging* 2004;**25**(5):589−97.

41. D'Andrea MR, Nagele RG, Wang H-Y, Peterson PA, Lee DHS. Evidence that neurones accumulating amyloid can undergo lysis to form amyloid plaques in AD. *Histopathology* 2001;**38**:120−34.

42. Gouras GK, Tsai J, Näslund J, Vincent B, Edgar M, Checler F, et al. Intraneuronal Aβ42 accumulation in human brain. *Am J Path* 2000;**156**(1):15−20.

43. Lord A, Kalimo H, Eckmang C, Zhang X-Q, Lannfelt L, Nilsson LGN. The Arctic Alzheimer mutation facilitates early intraneuronal Aβ aggregation and senile plaque formation in transgenic mice. *Neurobiol Aging* 2006;**27**:67−77.

44. D'Andrea MR. *Bursting neurons and fading memories*. New York: Elsevier Press; 2014.

45. Ahoa L, Pikkarainena M, Hiltunena M, Leinonenb V, Alafuzoff I. Immunohistochemical visualization of amyloid-β protein precursor and amyloid-β in extra- and intracellular compartments in the human brain. *J Alzheimers Dis* 2010;**20**:1015−28.

46. Tu S, Okamoto S-I, Lipton SA, Xu H. Review. Oligomeric Aβ-induced synaptic dysfunction in Alzheimer's disease. *Mol Neurodegener* 2014;**9**:48.

47. Takahashi RH, Milner TA, Li F, Nam EE, Edgar MA, Yamaguchi H, et al. Intraneuronal Alzheimer Aβ42 accumulates in multivesicular bodies and is associated with synaptic pathology. *Am J Pathol* 2002;**161**:1869−79.

48. Kimura N, Yanagisawa K, Terao K, Ono F, Sakakibara I, Ishii Y, et al. Age-related changes of intracellular Aβ in cynomolgus monkey brains. *Neuropathol Appl Neurobiol* 2005;**31**:170−80.

49. Kakio A, Nishimoto SI, Yanagisawa K, Kozutsumi Y, Matsuzaka K. Cholesterol-dependent formation of GM1 ganglioside-bound amyloid-protein, an endogenous seed for Alzheimer amyloid. *J Biol Chem* 2001;**276**:24985−90.

50. D'Andrea MR, Nagele RG, Wang H-Y, Lee DHS. Consistent immunohistochemical detection of intracellular β-amyloid42 in pyramidal neurons of Alzheimer's disease entorhinal cortex. *Neurosci Lett* 2002;**333**:163−6.

51. D'Andrea MR, Reiser PA, Polkovitch DA, Branchide B, Hertzog BH, Belkowski S, et al. The use of formic acid to embellish amyloid plaque detection in Alzheimer's disease tissues misguides key observations. *Neurosci Lett* 2003;**342**:114−18.

52. Ohyagi Y, Tsuruta Y, Motomura K, Miyoshi K, Kikuchi H, Iwaki T, et al. Intraneuronal amyloid β42 enhanced by heating but counteracted by formic acid. *J Neurosci Methods* 2007;**159**(1):134−8.

53. Christensen DZ, Bayer TA, Wirths O. Formic acid is essential for immunohistochemical detection of aggregated intraneuronal Aβ peptides in mouse models of Alzheimer's disease. *Brain Res* 2009;**1301**:116−25.

54. Cuello AC. Review. Intracellular and extracellular Aβ, a tale of two neuropathologies. *Brain Pathol* 2005;**15**:L66−71.

INTRACELLULAR CONSEQUENCES OF AMYLOID IN ALZHEIMER'S DISEASE

Origin(s) of Intraneuronal Amyloid

Aβ peptides accumulate within neurons. How they accumulate is a question of central importance to the understanding of AD pathogenesis. Either the origin of the intraneuronal Aβ is from an influx of exogenous Aβ42, and/or is derived from within the neuron through de novo Aβ42 synthesis, a defective transport system, abnormal clearance of Aβ as a result of defective proteolytic systems, endosomal and lysosomal changes, and/or various combinations of the above possibilities.[1–5]

In the brain, two pools of Aβ exist: intracellular and extracellular, and a dynamic relationship exists between them.[1] For example, in cells overexpressing the functional low-density lipoprotein receptor-related protein (LRP) minireceptor, a family of multiligand receptors known to have a high endocytosis rate and are highly expressed in the neurons of the brain, levels of intraneuronal Aβ42 increased as the levels of extracellular Aβ42 decreased.[6]

Intracellular Consequences of Amyloid in Alzheimer's Disease.
DOI: http://dx.doi.org/10.1016/B978-0-12-804256-4.00002-4

Aβ FROM WITHIN THE NEURON

Some of the earliest studies determined that the neuron itself was the source of the intraneuronal Aβ.[7-9] This was the belief for many years, and it explained how senile plaques were formed by the secretion of neuronal Aβ into the extracellular space, as both Aβ peptides, Aβ40 and Aβ42, are produced intracellularly prior to secretion.[10] Additional studies demonstrated de novo synthesis of Aβ peptides in cultured cells that accumulate gradually through slow intraneuronal production in the endoplasmic reticulum (ER) and Golgi.[10,11] In fact, aberrant Aβ catabolism within the neuronal cell body was considered among the key initial events in a series of pathological changes leading to AD.[12-14]

In another study, the intraneuronal sites that generate several Aβ species were characterized, as well as which Aβ species contributes to the secreted pool of peptides, and which were retained intraneuronally.[15] These studies suggested that the trans-Golgi network is the major, if not exclusive, intracellular compartment within which the γ-secretase is active.[15,16] Also, a population of insoluble Aβx-42 generated within the ER cannot be secreted, thereby explaining the presence of intraneuronal Aβ42, but suggested that the source of the plaque amyloid may not be from the neuron. However, this belief is contrary to many reports (see Chapter 5). In a 2006 study using an immunotherapy approach, extracellular Aβ was cleared before intracellular Aβ. After the antibody dissipated, the accumulation of intraneuronal Aβ suggested that intraneuronal Aβ may serve as a source for some of the extracellular amyloid deposits.[17]

The recycling of APP from the cell surface through the endocytic pathways also contributes to the generation of intraneuronal Aβ. One study investigated the genetics of several vacuolar protein sorting genes in AD and found that genetic variants in the neuronal sortilin-related receptor 1 (SORL1) led to reduced function that increased sorting of APP into Aβ-generating compartments suggesting that SORL1 directs trafficking of APP into the recycling pathways.[18] In the absence of SORL1, APP is released into late endosomal pathways where it is subjected to β- then γ-secretase cleavage that generate Aβ. Hence, intraneuronal accumulation can be dependent upon internal sorting mechanisms.

Under normal conditions of synaptic activity, intracellular and secreted Aβ are efficiently cleared, but clearance mechanisms become impaired with aging. Neprilysin, a zinc peptidase that degrade Aβ, declines with age.[19] Since neprilysin is localized to synapses, its age-related decline could explain the synaptic accumulation of Aβ42 that was observed both in AD transgenic mice and human AD brains. With progressive pathology, this accumulated intraneuronal Aβ42 could then be released to the extracellular space at high concentrations and become toxic to surrounding synapses.

Aβ FROM OUTSIDE THE NEURON

For all of the studies that suggest the origin of the intraneuronal Aβ is from within the neurons, others have looked outside the neuron for sources of the intraneuronal Aβ. Sources of the intraneuronal Aβ could arise from other cells in the brain, and/or from cells outside of the brain; both of which require a neuronal mechanism of Aβ uptake.

Aβ From Within the Brain

Secreted forms of Aβ from cells in the brain (eg, neurons, glia) represent other possible sources of the Aβ detected in neurons. Contributions of extraneuronal concentrations of Aβ in the brain could also originate from astrocytes since they also can secrete high levels of Aβ.[20] During early AD pathogenesis, it was suggested that Aβ initiates a vicious cycle of Aβ generation between astrocytes and neurons leading to chronic, sustained, and progressive neuroinflammation.[21] Other studies show that cytokines as well as neuronal-secreted oligomeric and fibrillar Aβ42 stimulate the production of APP processing in astrocytes, and when considering that the ratio of astrocytes to neurons is 5:1, astrocytes may actually represent significant sources of Aβ in the brain during the processes of neuroinflammation.[22] Astrocytes, like neurons, express the α7 nicotinic acetylcholine (α7) receptor, which is a calcium channel that has a very high affinity for Aβ (described later in the chapter).[23,24] The presence of an Aβ receptor on neurons and astrocytes could not only indicate receptor activation and signaling but also provide a mechanism of Aβ uptake.

Aβ From Outside the Brain

Peripheral Aβ could very well contribute to the high concentrations of Aβ in the brain, and subsequently into the neurons, especially in pathological conditions such as AD. The amount of Aβ mRNA using in situ hybridization was generally reduced in the neurons of AD brain tissues suggesting that the reported high levels of Aβ in the brain must had not originated from the neurons.[25]

The prevalence of uniformly small, Aβ42-positive granules in pyramidal neurons in brain regions with low plaque density of AD patients and in normal, age-matched, nondemented, control brain tissues may represent earlier stages of AD pathogenesis, which favors an entry mechanism for Aβ42 into these pyramidal neurons that includes endocytosis via a specific pathway(s) that is located on the membranes of those affected neurons.[2,3,26−29] Such a mechanism would require the Aβ42 to have some selective affinity for pyramidal neurons.

Aβ deposition in the brain could be the consequence of a breakdown of the blood−brain barrier (BBB, discussed later in the chapter), implying a peripheral origin of Aβ.[30,31] However, that possibility seemed improbable because it would imply a selective breakdown of the BBB especially since Aβ plaques were not detected in peripheral tissues in AD.[32] Another study also agreed because high plasma levels of Aβ in demented patients with Down syndrome, as compared to nondemented adults with Down syndrome, were unlikely related to the deposition of Aβ42 in the brain.[33]

Nevertheless, Aβ peptides are generated outside of the central nervous system (CNS) in appreciable quantities by the skeletal muscle, platelets, and vascular walls.[34−36] Other nonneural tissues expressing Aβ include: pancreas, kidney, spleen, heart, liver, testis, aorta, lung, intestines, skin as well as the adrenal, salivary, and thyroid glands.[37] These distinct reservoirs may allow for an active and dynamic interchange of Aβ peptides between the brain and periphery. These sources undoubtedly contribute to the pool of circulating Aβ. Therefore, a peripheral contribution of Aβ into the neurons of the brain is more and more possible. Also, small amounts of synthetic Aβ have been detected in the brain after intravenous injection of this peptide in rodents and primates.[31]

Plasma Aβ

If the source of the intraneuronal Aβ is from outside the brain, then Aβ must be present in the plasma. Studies detecting differences in the Aβ plasma levels between patients with AD and nondemented controls are inconsistent with limited clinical value, which could also be attributed to natural fluctuations due to diet, medications, stress, circadian rhythm, metabolic conditions, etc., and may be dependent upon the sensitivity of the assay conditions.[38−41] One group developed an immunomagnetic reduction assay to assess plasma levels of Aβ in AD and control patients.[38] Significant differences ($P = 0.001$) in Aβ42 plasma levels were observed between the AD and control patients, with 85.3% sensitivity and 88.5% specificity when using a cut-off value of 16.0 pg/mL.

Another study measured Aβ40 and Aβ42 in paired cerebrospinal fluid (CSF) and plasma samples from patients with AD, mild cognitive impairment (MCI, considered a very early stage of AD), and healthy control subjects, and observed a clear correlation between CSF and plasma levels for both Aβ40 and Aβ42 in healthy individuals, whereas no such correlation was observed for AD or MCI cases.[42] Similar to other studies, low levels of Aβ42 were detected in AD CSF, whereas there were no significant differences in plasma Aβ levels between the diagnostic groups. The findings from this study suggest that the normal equilibrium between CSF and plasma Aβ may be disrupted with the initiation of amyloid deposition in the brain.

Another study compared plasma levels of Aβ42 in 146 sporadic AD patients, 89 patients with MCI, and 89 age-matched controls. The results showed significantly lower levels of plasma Aβ in the AD that were unrelated to severity of the disease as assessed by mini-mental state examination (MMSE) score, age, sex, or apolipoprotein E (*APOE4*) status.[43] As a follow-up, 20 patients investigated at two time points 18 months apart did not demonstrate further decreases. Therefore, the reduction in Aβ42 may be a marker for AD status, specifically, a transition from normal status or MCI to AD, rather than a marker for neurodegenerative processes occurring in the disease.

Using a sandwich enzyme-linked immunosorbent assay technology, the association between plasma Aβ concentrations and the risk of dementia was assessed using Cox proportional hazard models.[44] Plasma Aβ42 concentration and the Aβ42/Aβ40 ratio at baseline were significantly decreased in the MCI patients who developed AD as compared to cognitively stable MCI patients. The baseline concentrations of Aβ40 were similar in all MCI groups. The Aβ42/Aβ40 ratio was superior to Aβ42 concentration with regard to identify incipient AD in MCI. The ratio of Aβ42 to Aβ40 rather than absolute levels of the peptides can aid in the identification of incipient AD among MCI patients.

Longitudinal studies of plasma Aβ40 and Aβ42 levels demonstrated wide temporal variation within and among the individuals tested.[39] Two independent plasma analyses showed virtually equivalent levels for total Aβ peptides: longitudinal = 517 pg/mL and therapeutic = 497 pg/mL. In both investigations there were no statistical correlations noted between Aβ values and MMSE scores, diagnoses, age, or gender. In support of the observations are several previous biomarker studies in which the plasma Aβ levels were not correlated with the diagnosis, medications, or with *APOE* genotype.

Conversely, as compared to cognitively normal age-matched subjects, significantly high plasma levels of Aβ were detected in women with MCI.[45,46] The high plasma concentrations of Aβ in these women with MCI were linked to the overproduction of this peptide with gender-dependent occurrence of sporadic late-onset AD and supported the findings that cellular overproduction coincided with gene mutations (eg, APP, presenilin 1, presenilin 2) in familial AD (FAD).[47,48] Interestingly, the plasma levels of Aβ42, a reported toxic fragment of Aβ, decreased during the course of the disease; hence, the elevated plasma Aβ levels were detected only at the preclinical stage that became similar to control plasma Aβ42 concentration levels when the patient was diagnosed as probable AD, and to those with definite AD.[47,49,50]

If lower plasma levels of Aβ in AD patients as compared to controls are real, then how can that be explained? One interpretation suggests that senile plaques sequester the Aβ thereby leading to lower peripheral concentrations.[51] Another suggests that reduced plasma levels of Aβ were attributed to the degeneration of neurons.[52] Another possibility could be due to vascular leakage of Aβ into the brain.[3] Once the Aβ escapes the vasculature to enter the brain, through a dysfunctional BBB, the neurons internalize the Aβ from the extracellular pool. Intracellular Aβ was always accompanied by increased extracellular Aβ, while in subjects without increased extracellular Aβ, there was no detection of intracellular Aβ.

Platelets

Among the many peripheral cells expressing forms of APP, platelets are particularly interesting since they show concentrations of the APP derivatives equivalent to those found in brains.[37,53–55] Platelets are considered the primary source of circulating amyloid (>95%) as they not only contain the full-length APP, but also sAPP and Aβ-containing carboxyl terminal fragments; hence, platelets possess the proteolytic machinery to produce Aβ and fragments similar to those produced in neurons.[35,54–58] Even though platelets process APP mostly through the α-secretase pathway to release sAPP, they do produce small amounts of Aβ predominantly Aβ40 over Aβ42 thereby providing evidence for β- and γ-secretase activities as well. A small proportion of full-length APP is present at the platelet surface, and this increases by threefold upon platelet activation with thrombin or collagen, and released Aβ from the platelets can in turn activate platelets.[35,54] Aβ can induce strong aggregation of human platelets, granule release, and integrin activation, similar to that elicited by physiological agonists thereby resulting in additional release of Aβ into the bloodstream.[58] These data suggest that APP is stored in secretory granules, and that it may have a physiological role in normal platelet function such as in the roles of platelet aggregation and coagulation, or in the repair mechanisms associated with injury.[35]

Epithelial Cells

Aβ and APP, its precursor protein, are present in the columnar absorptive epithelial cells of the small intestine.[59,60] A transient increase in the plasma concentration of APP follows the ingestion of dietary fats, and high-fat diet in animals can induce cerebral amyloidosis. Intracellular Aβ expression was observed in the rough ER and Golgi apparatus further suggesting that the cells were producing the Aβ.

The epithelial cells in the thyroid can also produce large amounts of APP and its APP fragments in comparison to other cell types such as lymphocytes and fibroblasts.[61] Although the function of APP and its fragments in thyroid function is not clear, it is believed that the APP fragments are released into circulation after thyroid cell damage.

The Blood-Brain Barrier

It is estimated that as many as 10 billion capillaries perfuse the human brain, and it is the BBB that separates what is inside these vessels from the brain (Fig. 2.1).[62] By definition, the BBB is a highly selective permeable barrier that separates the circulating blood from the brain chiefly by three components: the tight junctions that connect neighboring endothelial cells, a thick basement membrane, and the endfeet of the astrocytes in the brain. The barrier is selective because some gases (eg, oxygen, carbon dioxide), and lipid-soluble molecules can pass by passive diffusion, while other molecules such as glucose and amino acids can only pass by selective transport. The principal function of the barrier is to prevent large molecules (eg, bacteria) that could be toxic from entering the brain.[62,63]

When the BBB is impaired or compromised, unregulated vascular elements can enter the brain. For example, water regulation, a critical function of the BBB, is disturbed in AD leading to abnormal permeability and rates of water exchange across vessel walls.[63] Many studies report that the BBB is impaired as the result of aging or vascular alterations observed in neurodegenerative disorders (Fig. 2.1).[31,64–66] Other risk factors for AD, such as head trauma, stress, infection, hypertension, atherosclerosis, and stroke are known to disturb the integrity of BBB and may allow the passage of circulating Aβ into the brain.[64,67–70] Even in the absence of physical injury, inflammatory signals effect disruption of the BBB by activating endothelial cells, which not only allows previously restricted molecules (eg, antibiotics, phagocytes) from entering the brain, but unfortunately allows bacteria and viruses as well as other vascular-derived components into the brain.[71,72]

In addition to neuronal cell injury, AD- and Aβ-related cerebral vascular disorders are associated with an increased deposition of Aβ in brain parenchyma, and cerebral vasculature suggesting that damage to the cerebral endothelium and therefore the BBB may be involved in the development of AD (see Chapter 8).[62,65,73,74]

Genetic polymorphisms in APOE are a major risk factor for AD, and cause BBB disruption in animal models as proinflammatory cytokine secretion by pericytes were the initial insult in this process, which then was sustained by penetration of proinflammatory circulating factors.[75]

Fig. 2.2 shows the presence of extracellular Aβ42 in the AD brain in areas surrounding vessels thereby providing morphological evidence that Aβ42 can enter the brain through the vasculature.[76] Furthermore, Aβ42 deposition was found in the brains of rats receiving chronic intravenous injections of Aβ42 after disruption of the BBB by an experimentally administered stroke.[77]

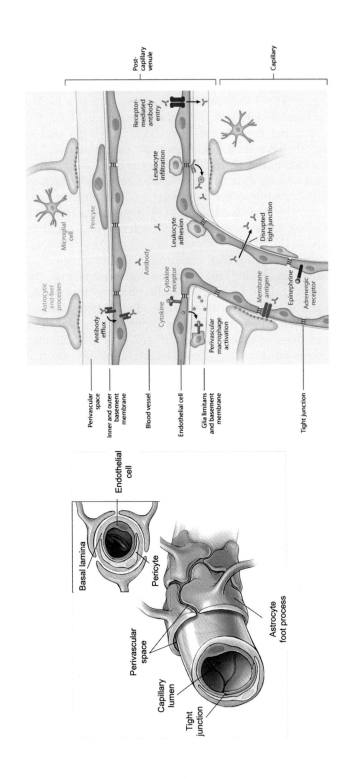

Post-capillary venule

Capillary

Astrocytic end-feet processes

Microglial cell

Pericyte

Antibody

Receptor-mediated antibody entry

Leukocyte infiltration

Leukocyte adhesion

Disrupted tight junction

Antibody efflux

Cytokine

Cytokine receptor

Membrane antigen

Epinephrine

Adrenergic receptor

Perivascular macrophage activation

Perivascular space

Inner and outer basement membrane

Blood vessel

Endothelial cell

Glia limitans and basement membrane

Tight junction

Basal lamina

Endothelial cell

Pericyte

Perivascular space

Capillary lumen

Tight junction

Astrocyte foot process

FIGURE 2.1 Left—Schematic diagram of the blood–brain barrier. Interstitial fluid flows along perivascular drainage pathways defined by the abluminal capillary surface and astrocyte end-feet, which ensheathes the vessels. Right—Diverse mechanisms of antibody entry into the CNS at the neurovascular interface. Anatomy and cell types: the neurovascular network forms a dense and intricate organ system in which numerous specialized cell types cooperate to form distinct classes of blood vessels with unique functions in regulating antibody access to the CNS parenchyma. The immediate barrier for preventing antibody entry into the CNS is the BBB formed by CNS endothelial cells. These cells form tight junctions that restrict paracellular diffusion of soluble factors and exhibit limited transcytosis and a relatively limited propensity for leukocyte adhesion under resting conditions. Astrocytes surround the basal surfaces of vessels with their foot processes and basement membranes. Most of the neurovascular surface area is composed of capillary microvessels, in which exchange of soluble factors occurs and in which astrocyte-derived membranes are opposed to endothelial-derived basement membranes. In larger postcapillary venules, a distinct perivascular space is observed between the astrocytic end-feet and the endothelial wall, where resident cells include pericytes (which induce barrier properties) and perivascular macrophages (which are thought to transduce inflammatory signals). Possible mechanisms of antibody entry: under normal physiological conditions, entry of immunoglobulins into the CNS is negligible. This restriction may be disrupted or bypassed under certain pathologic circumstances. Activation of CNS endothelial cells can result from activation of luminal receptors, including Toll-like receptors, cytokine receptors, and hormone receptors. Activation of perivascular macrophages might result from endothelial cell activation, which can further enhance disruption of tight junctions, permitting paracellular diffusion, increasing transcellular transport of soluble factors, and upregulating leukocyte adhesion molecules. Leukocyte adhesion to activated endothelial cells is likely to amplify relevant inflammatory signals, even in the absence of overt transmigration into the CNS. These processes are more likely to occur in brain microvascular capillaries. CNS inflammation may also cause inside-out disruption of the BBB. Antibodies with specificity for cell-surface molecules expressed by endothelial cells or perhaps by other neurovascular cells may enter the CNS by antigen binding and engaging transcytotic trafficking through receptor-mediated antibody entry. Leukocyte infiltration into the CNS may be counteracted by antibody efflux affected by polarized FcRn on endothelial cells. *BBB*, blood–brain barrier; *CNS*, central nervous system; *FcRn*, neonatal Fc receptor for immunoglobulin. *Source: Used with permission from Cardiovasc Psychiatry Neurol 2011;615829:9 pp. (left); Annu Rev Immunol 2013;31:345–85 (right).*

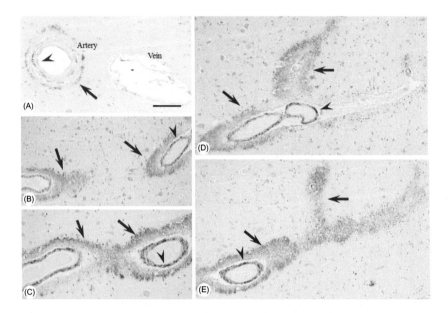

FIGURE 2.2 Aβ42-positive, vascular, diffuse plaques associated with an artery (*arrow*) and but not with a vein (A). Vascular-associated Aβ42 resembling diffuse plaques (*arrows*) in serial sectioning sets (B and C, D and E) clearly show its association with the vessels (B−E). Also note detection of intracellular Aβ42 in vascular smooth muscle cells (*arrowheads*). *Source: Used with permission from* Biotechnol Histochem 2010;85(5):133−47.

In some of the consequences of neuroinflammation in the AD brain, which can be triggered by Aβ, the microglia and astrocytes become activated. Among the many secreted factors from these reactive cells, vascular endothelial growth factor, a potent edemogenic agent, can lead to the breakdown of the BBB (see Chapter 7).[78]

Additional evidence that the BBB is allowing unregulated blood components into the brain comes from the detection of antibodies in the brain.[79−84] Antibodies can only penetrate the brain during development or under pathologic conditions.[64] The detection of autoantibodies within brain parenchyma suggests loss of the BBB integrity in AD patients.[85] In another example, autoantibodies were detected in aged nondemented human brains as well as in the AD brain (Fig. 2.3). Using computer-assisted image analysis, about a sevenfold increase in immunoglobulin immunolabeling was detected in brain tissues of AD brains as compared with age-matched, nondemented control brains (55% vs 8%, respectively).[81]

Patients with lupus produce anti-DNA antibodies that cross-react with the N-methyl-D-aspartate (NMDA) receptors and are capable of mediating excitotoxic neuronal death.[86] Mice induced by antigen to

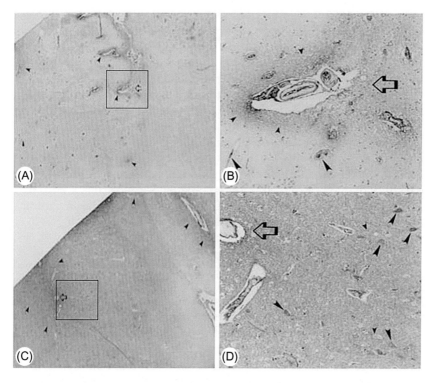

FIGURE 2.3   Representative images include Ig immunolabeling in "control" (A and B) and AD (C and D) cortical tissues at low and high magnifications. Areas in the black boxes in the lower-magnification micrographs (A and C) indicate areas of higher magnifications (B and D). Panels A and C show areas of Ig immunoreactivity (*arrowheads*) in the brain, which appears to clearly originate from vessels in the "control" tissues (A and B). At higher magnification of the "control" brain, panel B shows areas of Ig immunolabeling (*small arrowheads*) around cerebral vessels and areas of Ig (*large arrowheads*) in nearby vessels that do not show percolating Ig immunoreactivity in the neighboring brain parenchyma. Panel D identifies Ig-positive neurons (*large arrowheads*) in the AD tissue among Ig-negative neurons (*small arrowheads*). Ig, immunoglobulin. *Source: Used with permission from Brain Res 2003;982:19–30.*

express these anti-DNA antibodies have no neuronal damage until breakdown of the BBB occurs, thereby resulting in cognitive dysfunction, and altered hippocampal metabolism that can be reversed with memantine, as NMDA receptor antagonist. These results highlight the importance of a functional BBB to prevent CNS disease.

It is possible that the opposite is occurring, that soluble Aβ (sAβ) in the interstitial fluid of the brain may be removed via periarterial spaces, which drain into the cervical lymphatic nodes and finally into the venous circulation.[65,87] This bidirectional movement of Aβ through the BBB has been described as the dominant pathway regulating the concentrations of Aβ in the CNS under physiological and pathophysiological conditions and due to the soluble forms of Aβ in the plasma and within (or entrapped) the junctions of the endothelial cells that form the BBB.[65,88–90] The data in Fig. 2.2 are contrary to this possibility as the Aβ detected is associated with the arterial and not the venous system. Also, the higher hydrostatic pressure within the arterioles as compared with capillaries and venules will tend to push vascular elements out of the vessels, while the venous tends to be a passive vascular drain.

The tight junctions between the endothelial cells that help to maintain the integrity of the BBB can also be affected by plasma Aβ42.[65] Aβ peptide effects on cultured endothelial cells have been reported, which include induction of a central immunoregulatory molecule, CD40, phospholipid turnover, and overt cell death.[91,92] The disruption in claudin-5, an integral membrane protein of the tight-junction strands, at cell-to-cell contact sites of Aβ42-treated brain microvessel endothelial cells was time-dependent, already being evident after 24 h. This effect of Aβ peptide appeared to be specific to Aβ42, as neither the Aβ25–35 fragment nor Aβ1–40[Gln22] induced changes in claudin-5 localization.[65,73,93] These data suggested that Aβ42 effects on tight-junction protein complexes may alter BBB integrity, and contribute to the neuropathology in AD.

If the BBB is not able to function properly, just about anything from the vasculature could enter the brain. Apolipoprotein B (ApoB), the main apolipoprotein that transports cholesterol to peripheral tissues, is not present in normal brain but is present in AD brain.[94–96] Thus, when the BBB is leaky, as occurs in AD, circulating ApoB-containing cholesterol enters the brain, and is transported to neuronal endolysosomes via the same mechanisms for ApoE cholesterol as observed in a cholesterol-fed rabbit model of sporadic AD.[97]

Entry of Extracellular Aβ Into the Neurons

If the source of the intraneuronal Aβ is from outside the neurons, regardless of if from other cells in the brain, or from a peripheral source via vascular system, then how does the Aβ gain entry into the neurons?

A large number of intracellular, extracellular, and membrane-bound biomolecules (acute phase proteins, complement proteins, apolipoproteins, basement membrane proteins, membrane gangliosides/lipids, and various types of cell-surface receptors) have been shown to bind Aβ.[98–101] Binding causes changes in Aβ conformation, aggregation properties, and biological effects. Receptors capable of interacting with unmodified Aβ are the α7 receptors, Receptor for Advanced Glycosylation End Products (RAGE), macrophage scavenger receptors, formyl peptide receptor-like-1, NMDA receptor, insulin receptor, the

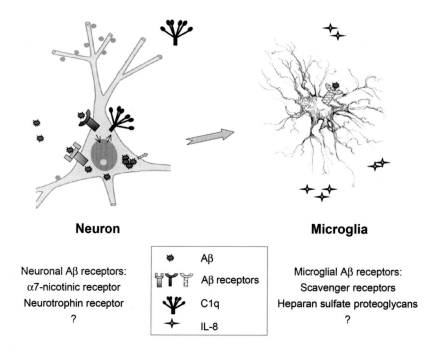

Neuron   **Microglia**

✳	Aβ
🌾🌾🌾	Aβ receptors
✲	C1q
✛	IL-8

Neuronal Aβ receptors:
α7-nicotinic receptor
Neurotrophin receptor
?

Microglial Aβ receptors:
Scavenger receptors
Heparan sulfate proteoglycans
?

FIGURE 2.4   Model of Aβ interaction with neurons and microglia in slice cultures. Exogenous Aβ peptide interacts with neuronal receptors leads to at least two separate consequences, in one of which C1q expression is upregulated in neurons. A second receptor mediates the secretion of certain modulatory molecules, which lead to microglial activation involving the expression of CD45, CR3, CD40, and IL-8. This does not exclude the direct interactions of Aβ with receptor(s) on microglia that may also contribute to microglial activation. *Aβ*, amyloid-β. *Source: Used with permission from* J Inflamm 2005;2:1.

P75 neurotrophin receptor (P75NTR) on neurons, integrin receptors on neurons and microglia, and heparan sulfate proteoglycans on microglia (Fig. 2.4, Table 2.1).[1,85,100−103] Endocytosis is a process that involves the coordination of many cellular proteins, which are mostly dependent on clathrin and dynamins to internalize molecules.[104] Clathrin is an adaptor protein that forms the primary component of the vesicle-coating complex, and dynamins are responsible for the membrane cleavage to release the invaginated vesicle from the plasma membrane. Endocytosis can also occur through clathrin-independent manner at the areas of the lipid rafts on the plasma membrane to form invaginations, called caveolae that are rich in cholesterol and sphingolipids, and are morphologically distinct from clathrin-coated invaginations.[105] It is possible that the internalization of Aβ can occur through both processes that may depend on the type of neuron as well as the area of the neuron.[103]

TABLE 2.1 Summary of Some of the Receptor Proteins Involved in the Neuronal Uptake of Aβ.

	Membrane receptors					Soluble receptors		
	α7nChR	AMPA	Integrins	LRP1	NMDA	RAGE	ApoE	RAP
Aβ binding	Yes	Yes	Yes	Via apoE	Probable	Yes	Yes	Yes
Intracellular colocalization with Aβ	Yes	Yes	No data	No data	No data	Yes	Yes	No data
Endocytic pathway	Clathrin-dependent and independent	Clathrin-dependent	Clathrin-dependent	Clathrin-dependent	Clathrin-dependent	No data	Clathrin-dependent, via LRP1	No data

α7nChR:α7, nicotinic acetylcholine receptor; AMPA, α-amino-3-hydroxy-5-methyl-4-isoxazolepropionic acid; ApoE, apolipoprotein E; LRP1, low-density lipoprotein receptor-related protein; NMDA, N-methyl-D-aspartate; RAGE, receptor for advanced glycosylation end products; RAP, receptor-associated protein.
Modified from Int J Alzheimer's Dis 2011;548380:11 pp.

α7 Receptors

The α7 receptors are one of the most widely expressed nicotinic receptors in the synaptic nervous system.[106–109] The α7 receptor is one of several nicotinic acetylcholine receptors that are ligand-gate ion channels whose opening is regulated in response to the binding of the neurotransmitter acetylcholine.[110] Such α7 receptors have an unusually high permeability to calcium, and regulate numerous calcium-dependent events in the nervous system, such as transmitter release, second messenger cascades, neurite extension, and both apoptosis and neuronal survival.[109] The receptors can also contribute directly to postsynaptic currents, and are expressed both at somato-dendritic and presynaptic sites on neurons in the hippocampus, a structure critical for memory formation, as activation of α7 receptors can promote long-term potentiation (LTP) at glutamatergic synapses.

Since these types of the cholinergic neurons are more prone to die in AD, the α7 receptor is one of the most studied examples of how Aβ gains entry into the neurons.[110,111] Up to 50% loss of neurons with choline acetyltransferase activity have been reported in the AD brains as compared with brains from subjects without AD.[110,112] In fact, one of the earlier hypotheses put forward about the cause of AD is based on deficits of nicotinic acetylcholine receptors and acetylcholine, and have led to therapies targeting cholinergic pathways to inhibit acetylcholine in activating enzymes.[113–115] The α7 receptor has a very high affinity (picomolar (pM) range) for Aβ42, which has been linked to mediating memory formation (see Chapter 3).[80] In the AD brain, the most vulnerable neurons are those containing the α7 receptor, and internalization of Aβ42 is facilitated by the high-affinity binding of Aβ42 to the α7 receptors on neuronal cell surfaces followed by endocytosis of the resulting complex, and its accumulation within the lysosomal compartment.[29]

Evidence shows that the α7 receptors can internalize the Aβ into neurons by endocytosis independent of clathrin and dynamin, a process that requires the polymerization of actin through activation of Ras-related C3 botulinum toxin substrate 1 (Rac1).[116] Aβ peptides can block the function of the α7 receptors specifically, reversibly, and with high affinity.[109] The blockade is noncompetitive and is exerted through the N-terminal extracellular portion of the receptor. It is voltage-independent and does not appear to result from Aβ42 acting as an open channel blocker because receptor activation is not required for the inhibition. However, rat Aβ42 peptide prepared in nonaggregate form directly activates rat α7 receptors expressed by *Xenopus* oocytes.[117] This is a high-affinity interaction; concentrations as low as 100 fetamolar Aβ42 are capable of inducing inward currents. Aβ42 is competitively antagonized by methyllycaconitine and cross-desensitized by 4-OH-GTS-21, both α7 receptor-selective agents.

Part of the pathology elicited by Aβ in AD may be due to aberrant activation of α7 receptors.[117] In particular, as the α7 receptor is a ligand-gated ion channel highly permeable to calcium, sustained increases of presynaptic calcium of α7 receptors evoked by Aβ in AD could lead to dysregulation of calcium homeostasis, thereby providing a molecular mechanism for the cholinergic dysfunction that is a hallmark of AD.[106,118–120]

Several sets of in vitro experiments provided additional support that intracellular accumulation of Aβ42 in neurons is facilitated by the α7 receptor.[29] The fate of exogenous Aβ42 and its interaction with the α7 receptor was explored in vitro using cultured, α7-transfected neuroblastoma cells that express elevated levels of this receptor. The transfected cells exhibited rapid binding, internalization, and accumulation of exogenous Aβ42, but not amyloid Aβ40. The rate and extent of amyloid Aβ42 internalization was related directly to the α7 receptor protein level, since (1) the rate of Aβ42 accumulation was much lower in untransfected cells that express much lower levels of this receptor, and (2) internalization was respectively blocked by α-bungarotoxin, an α7 receptor antagonist. The internalization of Aβ42 in α7-transfected cells was also blocked by phenylarsine oxide, an inhibitor of endocytosis. Although binding of Aβ peptides to the α7 receptor may also alter calcium ion influx and acetylcholine release, these results point to an important role for the α7 receptor in facilitating the entry, and intraneuronal accumulation of Aβ42 via endocytosis.

It is also important to consider that the α7 receptor is not only widely expressed in the central and peripheral nervous systems, but is also expressed in several nonneuronal cells, such as endothelial cells, bronchial epithelial cells, skin keratinocytes, and vascular smooth muscle cells.[121–124] As further discussed in Chapter 3 all of these cell types also contain intracellular Aβ possibly due to the α7 receptor. The early presence of nicotinic acetylcholine receptor proteins and gene transcripts in development suggests an important role for these receptors in modulating dendritic outgrowth, establishment of neuronal connections, and synaptogenesis during development.[125]

One cannot forget that the α7 receptor is a nicotine receptor as well and interestingly enough, chronic nicotine in vivo and tobacco use is capable in reducing brain Aβ in elderly individuals.[126] The discovery that nicotine and its mimetics can protect neurons against Aβ toxicity is of interest, especially in view of the observation that nicotine also enhances cognition.[110,111] The protective effect is blocked by the nicotinic antagonists, dihydro-β-erythroidine and mecamylamine.[127] The action of nicotine may also involve a positive feedback since exposure to nicotine increases expression of these receptors in brain and in cultured cells, and nicotine has also been reported to promote arteriogenesis, mediate angiogenesis, DNA synthesis, and proliferation in vascular endothelial cells, and in vascular smooth muscle cells.[113–115,128,129] The mechanisms of action remain uncertain, and further investigation is warranted to establish if targeting drugs to specific nicotinic acetylcholine receptor subtypes can treat or prevent AD.

Integrins and NMDA Receptors

The integrins and NMDA receptors also regulate clathrin-mediated internalization of Aβ peptides at the neuronal synapses, and are associated with lipid rafts, which are microdomains of signaling and trafficking areas in the plasma membrane.[130–134] These two classes of receptors operate cooperatively to regulate the endocytosis of Aβ in the neurons.

In addition to regulating Aβ intake, integrins also modulate the neurotoxic effects of sAβ.[130] When the interaction of Aβ with integrins is disrupted, less Aβ uptake was detected.

Exposure of the NMDA receptor to Aβ oligomers is known to promote endocytosis of the receptor as well as other signaling events associated with the receptor trafficking. For example, Aβ42 uptake occurred in 42% of the cultured hippocampal slices incubated with Aβ42 peptide alone and more than 80% of the slices cotreated with integrin antagonists.[130] Selective NMDA receptor antagonist D-(−)-2-amino-5-phosphonovalerate completely blocked internalization of Aβ42 in the hippocampal neurons.

These NMDA receptors also mediate Aβ-induced neurodegeneration, disruption in axonal transport, and impairment of synaptic transmission.[135–137] Interrupting the signaling NMDA receptors with an antagonist (eg, memantine), or by a transient inactivation, protected the cells against neuronal degeneration and Aβ toxicity.[135,138–140] It is interesting to note that Aβ mediates, and promotes NMDA receptor endocytosis possibly via the α7 receptor.[141,142]

Receptor for Advanced Glycosylation End Products

Another receptor that binds to Aβ is RAGE, which is a multiligand cell-surface receptor expressed by neurons, and a variety of cells including the inflammatory cells of the brain (astrocytes and microglia) and vascular cells (endothelial cells, pericytes, and smooth muscle cells).[85] RAGE binds monomeric, oligomeric, and even fibrillar Aβ at the surface of neurons, and then cointernalizes with Aβ. Blockade of RAGE decreases Aβ uptake and Aβ toxicity.[143–146] Also, increases in RAGE expression are upregulated in AD brains as compared to nondemented controls.[85] Evidence from in vivo and in vitro studies supports the role of RAGE in the pathogenesis of AD. RAGE activation by Aβ produces various consequences to include perturbation of neuronal properties and functions, amplification of glial inflammatory responses, elevation of oxidative stress and amyloidosis, increased Aβ influx at the BBB and vascular dysfunction, and induction of autoantibodies, which can be specifically blocked with an antibody to RAGE.[143,145,147] RAGE is considered a primary transporter of Aβ across the BBB into the brain from the systemic circulation.[148]

Apolipoprotein E

ApoEs are a group of proteins located along the neuronal plasma membrane (other than the synapse) and are involved in many important processes related to AD such as neuronal signaling, APP trafficking, and Aβ production.[149–151] Although not a receptor, some ApoEs act as soluble chaperones for the hydrophobic Aβ peptides in areas of lipid rafts.[150] Transgenic AD mice deficient in ApoE show reduced Aβ accumulation suggesting a role for ApoE in promoting Aβ accumulation.[152] In support, intraneuronal Aβ is significantly decreased in brains of transgenic AD mice lacking ApoE.[6] The presence of one or two APOE4 alleles strongly correlates with an increased accumulation of intraneuronal Aβ.[101]

ApoE co-localizes with intracellular Aβ42 in the neurons Alzheimer's tissues using IHC suggesting that ApoE is internalized with Aβ.[4,103,153–155] Moreover, ApoE has been observed to target neurotransmitter receptors such as the α7 receptor suggesting that the internalization of Aβ by these receptors may result from ApoE-receptor binding than by Aβ receptor binding.[110] However, the evidence presented in the α7 section does not appear to support this possibility.

The cell-surface receptors most associated with ApoE are the LRPs.[156] LRP1 is required for Aβ endocytosis in many cell types including cortical neurons from Tg2576 mice, glioblastoma, and neuroblastoma cells, fibroblasts, human cerebrovascular cells, synaptosomes and dorsal root ganglion cells, and brain endothelial cell lines.[7,157–163] LRP1 can bind to APP at the cell surface, and then mediate endocytosis of Aβ to the lysosomes and can also modulate Aβ production.[156] The same endocytic mechanism is thought to mediate the uptake of Aβ. Significantly reduced levels of intraneuronal Aβ using an LRP1 antagonist, and other in vivo experiments found that AD mice overexpressing LRP also increased intraneuronal accumulation of Aβ.[6,161]

Overexpression of the LRP minireceptor enhanced intracellular Aβ42 in the lysosomes in PC12 cells that correlated with decreased levels of extracellular Aβ42 in the medium.[6] Studies in 3-month-old transgenic mice showed Aβ42 accumulating with ApoE forming complexes in neurons even before the detection of extracellular plaques. Also, levels of intraneuronal Aβ42 were dramatically lower in mice lacking ApoE, thereby demonstrating that ApoE can increase the endocytosis of Aβ42 via binding to LRP, and perhaps other members of the low-density lipoprotein receptor family. These data suggest that although increased endocytosis of ApoE:Aβ42 complexes likely leads to lower brain ApoE levels, the Aβ42 was not completely degraded following endocytosis and accumulates.

SUMMARY

The origin(s) of amyloid, or specifically Aβ, is of the utmost importance to the information presented in the book. As presented, there are a myriad of plausible extraneuronal sources of Aβ including other cells in the brain, as well as peripheral sources of Aβ that may gain entry into the brain through a dysfunctional BBB. How extraneuronal Aβ is internalized in neurons is also presented, and includes Aβ binding proteins, as well as a variety of Aβ binding receptors.

References

1. Mohamed A, de Chaves EP. Review. Aβ internalization by neurons and glia. *Int J Alzheimer's Dis* 2011;**127984**:17 pp.
2. D'Andrea MR, Nagele RG, Wang HY, Peterson PA, Lee DHS. Evidence for neuronal origin of amyloid plaques in Alzheimer's disease. *Histopathology* 2001;**38**:120−34.
3. D'Andrea MR. *Bursting neurons and fading memories*. New York: Elsevier Press; 2014.
4. Gouras GK, Tsai J, Näslund J, Vincent B, Edgar M, Checler F, et al. Intraneuronal Aβ42 accumulation in human brain. *Am J Pathol* 2000;**156**(1):15−20.
5. Wang H-Y, D'Andrea MR, Nagele RG. Cerebellar diffuse amyloid plaques are derived from dendritic Aβ42 accumulations in Purkinje cells. *Neurobiol Aging* 2002;**23**:213−23.
6. Zerbinatti CV, Wahrle SE, Kim H, Cam JA, Bales K, Paul SM, et al. Apolipoprotein E and low density lipoprotein receptor-related protein facilitate intraneuronal Aβ42 accumulation in amyloid model mice. *J Biol Chem* 2006;**281**(47):36180−6.
7. LaFerla FM, Green KN, Oddo S. Review. Intracellular amyloid-β in Alzheimer's disease. *Nat Rev Neurosci* 2007;**8**:499−509.
8. Saavedra L, Mohamed A, Ma V, Kar S, de Chaves EP. Internalization of beta-amyloid peptide by primary neurons in the absence of apolipoprotein E. *J Biol Chem* 2007;**282**(49):35722−32.
9. Bayer TA, Wirths O. Review. Intracellular accumulation of amyloid-β − a predictor for synaptic dysfunction and neuron loss in Alzheimer's disease. *Front Aging Neurosci* 2010;**2**(8):10.
10. Turner RS, Suzuki N, Chyung ASC, Younkin SC, Lee W. Amyloids β40 and β42 are generated intracellularly in cultured human neurons and their secretion increases with maturation. *J Biol Chem* 1996;**271**(15):8966−70.
11. Cook DG, Forman MS, Sung JC, Leight S, Kolson DL, Iwatsubo T, et al. Alzheimer's Aβ (1−42) is generated in the endoplasmic reticulum/intermediate compartment of NT2N cells. *Nat Med* 1997;**3**:1021−3.
12. Rosenblum WI. The presence, origin and significance of Aβ peptide in the cell bodies of neurons. *J Neuropathol Exp Neurol* 1999;**58**:575−81.
13. Wilson CA, Doms RW, Lee V. Intracellular APP processing and Aβ production in Alzheimer disease. *J Neuropathol Exp Neurol* 1999;**58**:787−94.

14. Yang AJ, Chandswangbhuvana D, Shu T, Henschen A, Glabe CG. Intracellular accumulation of insoluble, newly synthesized Aβn–42 in amyloid precursor protein-transfected cells that have been treated with Aβ1–42. *J Biol Chem* 1999;**274**: 20650–6.

15. Greenfield JP, Tsai J, Gouras GK, Hai B, Thinakaran G, Checler F, et al. Endoplasmic reticulum and trans-Golgi network generate distinct populations of Alzheimer β-amyloid peptides. *PNAS* 1999;**96**:742–7.

16. Xu H, Shields D. Prohormone processing in the trans-Golgi network: endoproteolytic cleavage of prosomatostatin and formation of nascent secretory vesicles in permeabilized cells. *J Cell Biol* 1993;**122**:1169–84.

17. Oddo S, Caccamo A, Smith IF, Green KN, LaFerla FM. A dynamic relationship between intracellular and extracellular pools of Aβ. *Am J Pathol* 2006;**168**(1):184–94.

18. Rogaeva E, Meng Y, Lee JH, Gu Y, Kawarai T, Zou F, et al. The neuronal sortilin-related receptor SORL1 is genetically associated with Alzheimer's disease. *Nat Genet* 2007;**39**(2):168–77.

19. Tampellini D, Rahman N, Gallo EF, Huang Z, Dumont M, Capetillo-Zarate E, et al. Synaptic activity reduces intraneuronal Aβ, promotes APP transport to synapses and protects against Aβ-related synaptic alterations. *J Neurosci* 2009;**29**(31):9704–13.

20. Busciglio J, Gabuzda BH, Matsudaira P, Yanker BA. Generation of β-amyloid in the secretory pathway in neuronal and nonneuronal cells. *PNAS* 1993;**90**:2092–6.

21. Li C, Zhao R, Gao K, Wei Z, Yin MY, Lau LT, et al. Astrocytes: implications for neuroinflammatory pathogenesis of Alzheimer's disease. *Curr Alzheimer Res* 2011;**8**(1):67–80.

22. Zhao J, O'Connor T, Vassar R. The contribution of activated astrocytes to Aβ production: Implications for Alzheimer's disease pathogenesis. *J Neuroinflam* 2011;**8**:150.

23. Nagele RG, D'Andrea MR, Lee H, Venkataraman V, Wang H-Y. Astrocytes accumulate Aβ42 and give rise to astrocytic amyloid plaques in Alzheimer disease brains. *Brain Res* 2003;**971**:197–209.

24. Yua W-F, Guana Z-Z, Bogdanovic N, Nordberg A. High selective expression of α7 nicotinic receptors on astrocytes in the brains of patients with sporadic Alzheimer's disease and patients carrying Swedish APP 670/671 mutation: a possible association with neuritic plaques. *Exp Neurol* 2005;**192**:215–25.

25. Shoji M, Kawarabayashi T, Matsubara E, Ikeda M, Ishiguro K, Harigaya Y, et al. Distribution of amyloid β protein precursor in the Alzheimer's disease brain. *Psychiatry Clin Neurosci* 2000;**54**:45–54.

26. D'Andrea MR, Lee DHS, Wang H-Y, Nagele RG. Targeting intracellular Aβ42 for Alzheimer's disease drug discovery. *Drug Dev Res* 2002;**56**:194–200.

27. Cataldo AM, Hamilton DJ, Nixon RA. Lysosomal abnormalities in degenerating neurons link neuronal compromise to senile plaque development in Alzheimer disease. *Brain Res* 1994;**640**:68–80.

28. Wang H-Y, Lee DHS, D'Andrea MR, Peterson PA, Shank RP, Reitz AB. Beta-amyloid1-42 binds to α 7 nicotinic acetylcholine receptor with high affinity: implications for Alzheimer's disease pathology. *J Biol Chem* 2000;**275**:5626–32.

29. Nagele RG, D'Andrea MR, Anderson WJ, Wang H-Y. Intracellular accumulation of β-amyloid$_{1\text{-}42}$ in neurons is facilitated by the α 7 nicotinic receptor in Alzheimer's disease. *Neurosci* 2002;**110**:199–211.

30. Glenner GG. Congophilic microangiopathy in the pathogenesis of Alzheimer's syndrome (presenile dementia). *Med Hypotheses* 1979;**5**:1231–6.

31. Kuo Y-M, Emmerling MR, Lampert HC, Hempelman SR, Kokjohn TA, Woods AS, et al. High levels of circulating Aβ42 are sequestered by plasma proteins in Alzheimer's disease. *Biochem Biophys Res Com* 1999;**257**:787–91.

32. Goedert M. Neuronal localization of amyloid beta protein precursor mRNA in normal human brain and in AD. *EMBO J* 1987;**6**(12):3627–32.

33. Schupf N, Patela B, Silverman W, Zigman WB, Zhong N, Tycko B, et al. Elevated plasma amyloid β-peptide 1-42 and onset of dementia in adults with Down syndrome. *Neurosci Lett* 2001;**301**:199−203.
34. Kuo YM, Kokjohn TA, Watson MD, Woods AS, Cotter RJ, Sue LI, et al. Elevated Aβ42 in skeletal muscle of Alzheimer disease patients suggests peripheral alterations of AβPP metabolism. *Am J Pathol* 2000;**156**:797−805.
35. Li QX, Whyte S, Tanner JE, Evin G, Beyreuther K, Masters CL. Secretion of Alzheimer's disease Aβ amyloid peptide by activated human platelets. *Lab Invest* 1998;**78**:461−9.
36. Van Nostrand WE, Melchor JP. Disruption of pathologic amyloid β-protein fibril assembly on the surface of cultured human cerebrovascular smooth muscle cells. *Amyloid* 2001;**8**(Suppl. 1):20−7.
37. Selkoe DJ, Podlisny MB, Joachim CL, Vickers EA, Lee G, Fritz LC, et al. Beta-amyloid precursor protein of Alzheimer disease occurs as 110- to 135-kilodalton membrane-associated proteins in neural and nonneural tissues. *PNAS* 1988;**85**:7341−5.
38. Chiu MJ, Yang SY, Chen TF, Chieh JJ, Huang TZ, Yip PK, et al. New assay for old markers-plasma β amyloid of mild cognitive impairment and Alzheimer's disease. *Curr Alzheimer Res* 2012;**9**(10):1142−8.
39. Roher AE, Esh CL, Kokjohn TA, Castaño EM, Van Vickle CD, Kalback WM, et al. Aβ peptides in human plasma and tissues and their significance for Alzheimer's disease. *Alzheimers Dement* 2009;**5**(1):18−29.
40. Dickstein DL, Walsh J, Brautigam H, Stockton Jr SD, Gandy S, Hof PR. Review. Role of vascular risk factors and vascular dysfunction in Alzheimer's disease. *Mt Sinai J Med* 2010;**77**:82−102.
41. Bateman RJ, Wen G, Morris JC, Holtzman DM. Fluctuations of CSF amyloid-β levels: implications for a diagnostic and therapeutic biomarker. *Neurology* 2007;**68**:666−9.
42. Giedraitis V, Sundelöfa J, Irizarry MC, Gårevikc N, Hyman BT, Wahlund L-O, et al. The normal equilibrium between CSF and plasma amyloid β levels is disrupted in Alzheimer's disease. *Neurosci Lett* 2007;**427**(3):127−31.
43. Pesaresi M, Lovati C, Bertora P, Mailland E, Galimberti D, Scarpini E, et al. Plasma levels of β-amyloid (1−42) in Alzheimer's disease and mild cognitive impairment. *Neurobiol Aging* 2006;**27**:904−5.
44. Fei M, Jianghua W, Rujuan M, Wei Z, Qian W. The relationship of plasma Aβ levels to dementia in aging individuals with mild cognitive impairment. *J Neurol Sci* 2011;**305** (1−2):92−6.
45. Petersen RC, Stevens JC, Ganguli M, Tangalos EG, Cummings JL, DeKosky ST. Practice parameter: early detection of dementia: mild cognitive impairment (an evidence-based review). Report of the Quality Standards Subcommittee of the American Academy of Neurology. *Neurology* 2001;**56**:1133−42.
46. Morris JC, Storandt M, Miller JP, et al. Mild cognitive impairment represents early-stage Alzheimer disease. *Arch Neurol* 2001;**58**:397−405.
47. Assini A, Cammarata S, Vitali A, Colucci M, Giliberto L, Borghi R, et al. Plasma levels of amyloid β-protein 42 are increased in women with mild cognitive impairment. *Neurology* 2004;**63**:828−31.
48. Scheuner D, Eckman C, Jensen M, et al. Secreted amyloid β-protein similar to that in the senile plaques of Alzheimer's disease is increased in vivo by the presenilin 1 and 2 and APP mutations linked to familial Alzheimer's disease. *Nat Med* 1996;**2**:864−70.
49. Tamaoka A, Fukushima T, Sawamura N, et al. Amyloid β-protein in plasma from patients with sporadic Alzheimer's disease. *J Neurol Sci* 1996;**141**:65−8.
50. Mehta PD, Pirttila T, Mehta SP, Sersen EA, Aisen PS, Wisniewski HM. Plasma and cerebrospinal fluid levels of amyloid β proteins 1-40 and 1-42 in Alzheimer disease. *Arch Neurol* 2000;**57**:100−5.

INTRACELLULAR CONSEQUENCES OF AMYLOID IN ALZHEIMER'S DISEASE

51. Sjogren M, Davidsson P, Wallin A, Granerus AK, Grundstrom E, Askmark H, et al. Decreased CSF-β-amyloid 42 in Alzheimer's disease and amyotrophic lateral sclerosis may reflect mismetabolism of β-amyloid induced by disparate mechanisms. *Dement Geriatr Cogn Disord* 2002;**13**:112–18.

52. Kawarabayashi T, Younkin LH, Saido TC, Shoji M, Ashe KH, Younkin SG. Age-dependent changes in brain, CSF, and plasma amyloid (β) protein in the Tg2576 transgenic mouse model of Alzheimer's disease. *J Neurosci* 2001;**21**:372–81.

53. Van Nostrand WE, Schmaier AH, Farrow JS, Cunningham DD. Protease nexin II (amyloid β protein precursor): a platelet α-granule protein. *Science* 1990;**248**:745–8.

54. Zhang W, Huang W, Jing F. Review. Contribution of blood platelets to vascular pathology in Alzheimer's disease. *J Blood Med* 2013;**7**:141–7.

55. Evin G, Li Q-X. Review. Platelets and Alzheimer's disease: potential of APP as a biomarker. *World J Psychiatr* 2012;**2**(6):102–13.

56. Cattabeni F, Colciaghi F, Di Luca M. Review. Platelets provide human tissue to unravel pathogenic mechanisms of Alzheimer disease. *Prog Neurol Psychopharmacol Biol Psychiatry* 2004;**28**:763–70.

57. Borroni B, Akkawi N, Martini G, Colciaghi F, Prometti P, Rozzini L, et al. Review. Microvascular damage and platelet abnormalities in early Alzheimer's disease. *J Neurol Sci* 2002;**203**:189–93.

58. Sonkar VK, Kulkarni PP, Dash D. Amyloid β peptide stimulates platelet activation through RhoA-dependent modulation of actomyosin organization. *FASEB J* 2014;**28**(4):1819–29.

59. Galloway S, Jian L, Johnsen R, Chew S, Mamo JCL. Beta-amyloid or its precursor protein is found in epithelial cells of the small intestine and is stimulated by high-fat feeding. *J Nutr Biochem* 2007;**18**:279–84.

60. Galloway S, Pallebage-Gamarallage M, Takechi R, Jian L, Johnsen RD, Dhaliwal SS, et al. Synergistic effects of high fat feeding and apolipoprotein E deletion on enterocytic amyloid-β abundance. *Lipids Health Dis* 2008;**7**:15.

61. Schmitt TL, Steiner E, Klingler P, Lassmann H, Grubeck-Loebenstein B. Thyroid epithelial cells produce large amounts of the Alzheimer β-amyloid precursor protein (APP) and generate potentially amyloidogenic APP fragments. *J Clin Endocrinol Metab* 1995;**80**(12):35113–19.

62. Zlokovic BV. Review. The blood-brain barrier in health and chronic neurodegenerative disorders. *Neuron* 2008;**57**:178–201.

63. Anderson VC, Lenar DP, Quinn JF, Rooney WD. Review. The blood-brain barrier and microvascular water exchange in Alzheimer's disease. *Cardiovasc Psychiatry Neurol* 2011;**615829**:9 pp.

64. Diamond B, Honig G, Mader S, Brimberg L, Volpe BT. Brain-reactive antibodies and disease. *Annu Rev Immunol* 2013;**31**:345–85.

65. Marco S, Skaper SD. Amyloid β-peptide1-42 alters tight junction protein distribution and expression in brain microvessel endothelial cells. *Neurosci Lett* 2006;**401**:219–24.

66. Donahue JE, Johanson CE, Apolipoprotein E. amyloid-β, and blood-brain barrier permeability in Alzheimer disease. *J Neuropathol Exp Neurol* 2008;**67**(4):261–70.

67. Van Duijn CM, Tanja TA, Haaxma R, Schulte W, Saan RJ, Lameris AJ, et al. Head trauma and the risk of Alzheimer's disease. *Am J Epidemiol* 1992;**135**:775–82.

68. Skoog I, Lernfelt B, Landahl S, Palmertz B, Andreasson LA, Nilsson L, et al. 15-year longitudinal study of blood pressure and dementia. *Lancet* 1996;**347**:1141–5.

69. Hofman A, Ott A, Breteler MM, Bots ML, Slooter AJ, van Harskamp F, et al. Atherosclerosis, apolipoprotein E, and prevalence of dementia and Alzheimer's disease in the Rotterdam Study. *Lancet* 1997;**349**:151–4.

70. Snowdon DA, Greiner LH, Mortimer JA, Riley KP, Greiner PA, Markesbery WR. Brain infarction and the clinical expression of Alzheimer disease. The Nun Study. *JAMA* 1997;**277**:813−17.

71. Lossinsky AS, Shivers RR. Review. Structural pathways for macromolecular and cellular transport across the blood-brain barrier during inflammatory conditions. *Histol Histopathol* 2004;**19**:535−64.

72. Dénes Á, Humphreys N, Lane TE, Grencis R, Rothwell N. Chronic systemic infection exacerbates ischemic brain damage via a CCL5 (regulated on activation, normal T-cell expressed and secreted)-mediated proinflammatory response in mice. *J Neurosci* 2010;**30**:10086−95.

73. Castaño EM, Prelli F, Soto C, Beavis R, Matsubara E, Shoji M, et al. The length of amyloid-β in hereditary cerebral hemorrhage with amyloidosis, Dutch type. Implications for the role of amyloid-β 1-42 in Alzheimer's disease. *J Biol Chem* 1996;**271**:32185−91.

74. Yamada M. Review. Cerebral amyloid angiopathy: an overview. *Neuropathology* 2000;**20**:8−22.

75. Bell RD, Winkler EA, Singh I, Sagare AP, Deane R, et al. Apolipoprotein E controls cerebrovascular integrity via cyclophilin A. *Nature* 2012;**485**:512−16.

76. D'Andrea MR, Nagele RG. Morphologically distinct types of amyloid plaques point the way to a better understanding of Alzheimer's disease pathogenesis. *Biotechnic Histochem* 2010;**85**(2):133−47.

77. Pluta R, Misicka A, Januszewski S, Barcikowska M, Lipkowski AW. Transport of human β-amyloid peptide through the rat blood-brain barrier after global cerebral ischemia. *Acta Neurochir* 1997;**S70**:247−9.

78. Salhia B, Angelov L, Roncari L, Wu X, Shannon P, Guhaa A. Expression of vascular endothelial growth factor by reactive astrocytes and associated neoangiogenesis. *Brain Res* 2000;**883**:87−97.

79. Sasaki N, Toki S, Chowei H, Saito T, Nakano N, Hayashi Y, et al. Immunohistochemical distribution of the receptor for advanced glycation end products in neurons and astrocytes in Alzheimer's disease. *Brain Res* 2001;**888**:256−62.

80. Garcia-Osta A, Alberini CM. Amyloid β mediates memory formation. *Learn Mem* 2009;**16**:267−72.

81. D'Andrea MR. Evidence linking neuronal cell death to autoimmunity in Alzheimer's disease. *Brain Res* 2003;**982**:19−30.

82. Racke MM, Boone LI, Hepburn DL, Parsadainian M, Bryan MT, Ness DK, et al. Exacerbation of cerebral amyloid angiopathy-associated microhemorrhage in amyloid precursor protein transgenic mice by immunotherapy is dependent on antibody recognition of deposited forms of amyloid β. *J Neurosci* 2005;**25**:629.

83. D'Andrea MR. Evidence the immunoglobulin-positive neurons in Alzheimer's disease are dying via the classical antibody-dependent complement pathway. *Am J Alzheimer's Dis Other Demen* 2005;**20**:144.

84. D'Andrea MR. Add Alzheimer's disease to the list of autoimmune diseases. *Med Hypotheses* 2005;**64**:458−63.

85. Lue L-F, Yan SD, Stern DM, Walker DG. Review. Preventing activation of receptor for advanced glycation endproducts in Alzheimer's disease. *Curr Drug Targets* 2005;**4**:249−66.

86. Kowa C, DeGiorgio LA, Nakaoka T, Hetherington H, Huerta PT, Diamond B, et al. Cognition and immunity: antibody impairs memory. *Immunity* 2004;**21**:179−88.

87. Weller RO, Massey A, Newman TA, Hutchings M, Kuo Y-M, Roher AE. Cerebral amyloid angiopathy: amyloid β accumulates in putative interstitial fluid drainage pathways in Alzheimer's disease. *Am J Pathol* 1998;**153**:725−33.

88. Zlokovic BV, Ghiso J, Mackic JB, McComb JG, Weiss MH, Frangione B. Blood–brain barrier transport of circulating Alzheimer's amyloid β. *Biochem Biophys Res Commun* 1993;**197**:1034–40.

89. Mackic JB, Bading J, Ghiso J, Walker L, Wisniewski T, Frangione B, et al. Circulating amyloid-β peptide crosses the blood–brain barrier in aged monkeys and contributes to Alzheimer's disease lesions. *Vascul Pharmacol* 2002;**38**:303–13.

90. Glenner GG, Wong CW. Alzheimer's disease and Down's syndrome: sharing of a unique cerebrovascular amyloid fibril protein. *Biochem Biophys Res Commun* 1984;**122**:1131–5.

91. Anfuso CD, Lupo G, Alberghina M. Amyloid β but not bradykinin induces phosphatidylcholine hydrolysis in immortalized rat brain endothelial cells. *Neurosci Lett* 1999;**271**:151–4.

92. Xu J, Chen SW, Ku G, Ahmed SH, Xu JM, Chen H, et al. Amyloid β peptide-induced cerebral endothelial cell death involve mitochondrial dysfunction and caspase activation. *J Cereb Blood Flow Metab* 2001;**21**:702–10.

93. Yankner BA, Duffy LK, Kirschner DA. Neurotrophic and neurotoxic effects of amyloid β protein: reversal by tachykinin neuropeptides. *Science* 1990;**250**:279–82.

94. Pitas RE, Boyles JK, Lee SH, Hui D, Weisgraber KH. Lipoproteins and their receptors in the central nervous system. Characterization of the lipoproteins in cerebrospinal fluid and identification of apolipoprotein B, E(LDL) receptors in the brain. *J Biol Chem* 1987;**262**:14352–60.

95. Namba Y, Tsuchiya H, Ikeda K. Apolipoprotein B immunoreactivity in senile plaque and vascular amyloids and neurofibrillary tangles in the brains of patients with Alzheimer's disease. *Neurosci Lett* 1992;**134**:264–6.

96. Takechi R, Galloway S, Pallebage-Gamarallage M, Wellington C, Johnsen R, Mamo JC. Three-dimensional colocalization analysis of plasma-derived apolipoprotein B with amyloid plaques in APP/PS1 transgenic mice. *Histochem Cell Biol* 2009;**131**:661–6.

97. Hui L, Chen X, Geiger JD. Endolysosome involvement in LDL cholesterol-induced Alzheimer's disease-like pathology in primary cultured neurons. *Life Sciences* 2002;**91**:1159–68.

98. Verdier Y, Zarandi M, Penke BJ. Amyloid β-peptide interactions with neuronal and glial cell plasma membrane: binding sites and implications for Alzheimer's disease. *Pept Sci* 2004;**10**:229.

99. Koenigsknecht J, Landreth GJ. Microglial phagocytosis of fibrillar β-amyloid through a β1 integrin-dependent mechanism. *Neurosci* 2004;**24**:9838.

100. Verdier Y, Penke B. Binding sites of amyloid β-peptide in cell plasma membrane and implications for Alzheimer's disease. *Curr Protein Pept Sci* 2004;**5**(1):19–31.

101. Christensen DZ, Schneuder-Axmann T, Lucassen PJ, Bayer TA, Wirths O. Accumulation of intraneuronal Aβ correlated with ApoE4 genotype. *Acta Neuropathol* 2010;**119**:555–66.

102. Fan R, Tenner AJ. Differential regulation of Aβ42-induced neuronal C1q synthesis and microglial activation. *J Neuroinflammation* 2005;**2**:1.

103. Lai AY, McLaurin J. Review. Mechanisms of amyloid-beta peptide uptake by neurons: The role of lipid rafts and lipid raft-associated proteins. *Int J Alzheimer's Dis* 2011;**548380**:11 pp.

104. Kumari S, Mg S, Mayor S. Endocytosis unplugged: multiple ways to enter the cell. *Cell Research* 2010;**20**(3):256–75.

105. Lajoie P, Nabi IR. Regulation of raft-dependent endocytosis. *J Cell Mol Med* 2007;**11**(4):644–53.

106. Dougherty JJ, Wu J, Nichols RA. Beta-amyloid regulation of presynaptic nicotinic receptors in rat hippocampus and neocortex. *J Neurosci* 2003;**23**(17):6740–7.

107. Sargent PB. Review. The diversity of neuronal nicotinic acetylcholine receptors. *Annu Rev Neurosci* 1993;**16**:403–33.
108. McGehee DS, Role LW. Physiological diversity of nicotinic acetylcholine receptors expressed by vertebrate neurons. *Annu Rev Physiol* 1995;**57**:521–46.
109. Liu Q-S, Kawai H, Berg DK. Beta-amyloid peptide blocks the response of α7-containing nicotinic receptors on hippocampal neurons. *PNAS* 2001;**98**(8):4734–9.
110. Buckingham SD, Jones AK, Brown LA, Sattelle DB. Review. Nicotinic acetylcholine receptor signalling: roles in Alzheimer's disease and amyloid neuroprotection. *Pharmacol Rev* 2009;**61**(1):39–61.
111. Kadir A, Almkvist O, Wall A, Långstrom B, Nordberg A. PET imaging of cortical 11C-nicotine binding correlates with the cognitive function of attention in Alzheimer's disease. *Psychopharmacology (Berl)* 2006;**188**:509–20.
112. Price DL. Review. New perspectives on Alzheimer's disease. *Annu Rev Neurosci* 1986;**9**:489–512.
113. Bartus RT, Dean III RL, Beer B, Lippa AS. The cholinergic hypothesis of geriatric memory dysfunction. *Science* 1982;**217**:408–14.
114. Francis PT, Palmer AM, Snape M, Wilcock GK. Review. The cholinergic hypothesis of Alzheimer's disease: a review of progress. *J Neurol Neurosurg Psychiatry* 1999;**66**:137–47.
115. Arneric SP, Holladay M, Williams M. Review. Neuronal nicotinic receptors: a perspective on two decades of drug discovery research. *Biochem Pharmacol* 2007;**74**:1092–101.
116. John PAS. Cellular trafficking of nicotinic acetylcholine receptors. *Acta Pharmacol Sin* 2009;**30**(6):656–62.
117. Dineley KT, Bell KA, Bui D, Sweatt JD. β-amyloid peptide activates α7 nicotinic acetylcholine receptors expressed in *Xenopus* oocytes. *J Biol Chem* 2002;**277**(28):25056–61.
118. Seguela P, Wadiche J, Dineley-Miller K, Dani JA, Patrick JW. Molecular cloning, functional properties, and distribution of rat brain α 7: a nicotinic cation channel highly permeable to calcium. *J Neurosci* 1993;**13**:596–604.
119. Bowen DM, Smith CB, White P, Davison AN. Neurotransmitter-related enzymes and indices of hypoxia in senile dementia and other abiotrophies. *Brain* 1976;**99**:459–96.
120. Davies P, Maloney AJ. Selective loss of central cholinergic neurons in Alzheimer's disease. *Lancet* 1976;**2**:1403.
121. Wang Y, Pereira EFR, Maus ADJ, Ostlie NS, Navaneetham D, Lei S, et al. Human bronchial epithelial and endothelial cells express a7 nicotinic acetylcholial receptors. *Mol Pharmacol* 2001;**60**(6):1201–9.
122. Li X-W, Wang H. Non-neuronal nicotinic alpha 7 receptor, a new endothelial target for revascularization. *Life Sci* 2006;**78**:1863–70.
123. Macklin KD, Mausad AD, Pereira EF, Albuquerque EX, Conti-Fine BM. Human vascular endothelial cells express functional nicotinic acetylcholine receptors. *J Pharmacol Exp Ther* 1998;**287**(1):435–9.
124. Villablanca AC. Nicotine stimulates DNA synthesis and proliferation in vascular endothelial cells in vitro. *J Appl Physiol* 1998;**84**(6):2089–98.
125. Schultz DW, Loring RH, Aizenman E, Zigmond RE. Autoradiographic localization of putative nicotinic receptors in the rat brain using ^{125}I-neuronal bungarotoxin. *J Neurosci* 1991;**11**(1):287–97.
126. Kummer C, Wehner S, Quast T, Werner S, Herzog V. Expression and potential function of β-amyloid precursor proteins during cutaneous wound repair. *Exp Cell Res* 2002;**280**(2):222–32.
127. Strong R, Huang JS, Huang SS, Chung HD, Hale C, Burke WJ. Degeneration of the cholinergic innervation of the locus ceruleus in Alzheimer's disease. *Brain Res* 1991;**542**:23–8.

128. Heeschen C, Weis M, Aicher A, Dimmeler S, Cooke JP. A novel angiogeneic pathway mediated by non-neuronal nicotinic acetylcholine receptors. *J Clin Invest* 2002; **110**(4):527−36.

129. Pestana IA, Vazquez-Padron RA, Aitouche A, Pham SM. Nicotinic and PDGF-receptor function are essential for nicotine-stimulated mitogenesis in human vascular smooth muscle cells. *J Cell Biochem* 2005;**96**:986−95.

130. Bi X, Gall CM, Zhou J, Lynch G. Uptake and pathogenic effects of amyloid beta peptide 1-42 are enhanced by integrin antagonists and blocked by NMDA receptor antagonists. *Neuroscience* 2002;**112**(4):827−40.

131. Hering H, Lin CC, Sheng M. Lipid rafts in the maintenance of synapses, dendritic spines, and surface AMPA receptor stability. *J Neurosci* 2003;**23**(8):3262−71.

132. Besshoh S, Bawa D, Teves L, Wallace MC, Gurd JW. Increased phosphorylation and redistribution of NMDA receptors between synaptic lipid rafts and post-synaptic densities following transient global ischemia in the rat brain. *J Neurochem* 2005;**93**(1):186−94.

133. Vassilieva EV, Gerner-Smidt K, Ivanov IA, Nusrat A. Lipid rafts mediate internalization of β-integrin in migrating intestinal epithelial cells. *Am J Physiol* 2008;**295**(5): G965−76.

134. <https://en.wikipedia.org/wiki/Lipid_raf>. [accessed 19.08.15].

135. Miguel-Hidalgo JJ, Alvarez XA, Cacabelos R, Quack G. Neuroprotection by memantine against neurodegeneration induced by β-amyloid (1−40). *Brain Res* 2002;**958** (1):210−21.

136. Harkany T, Abraham I, Timmerman W. Beta-amyloid neurotoxicity is mediated by a glutamate-triggered excitotoxic cascade in rat nucleus basalis. *Eur J Neurosci* 2000;**12** (8):2735−45.

137. Decker H, Lo KY, Unger SM, Ferreira ST, Silverman MA. Amyloid-β peptide oligomers disrupt axonal transport through an NMDA receptor-dependent mechanism that is mediated by glycogen synthase kinase 3β in primary cultured hippocampal neurons. *J Neurosci* 2010;**30**(27):9166−71.

138. Mattson MP, Cheng B, Davis D, Bryant K, Lieberburg I, Rydel RE. Beta-amyloid peptides destabilize calcium homeostasis and render human cortical neurons vulnerable to excitotoxicity. *J Neurosci* 1992;**12**(2):376−89.

139. Tremblay R, Chakravarthy B, Hewitt K, et al. Transient NMDA receptor inactivation provides long-term protection cultured cortical neurons from a variety of death signals. *J Neurosci* 2000;**20**(19):7183−92.

140. Song MS, Rauw G, Baker GB, Kar S. Memantine protects rat cortical cultured neurons against β-amyloid-induced toxicity by attenuating tau phosphorylation. *Eur J Neurosci* 2008;**28**(10):1989−2002.

141. Snyder EM, Nong Y, Almeida CG, et al. Regulation of NMDA receptor trafficking by amyloid-β. *Nat Neurosci* 2005;**8**(8):1051−8.

142. Kurup P, Zhang Y, Xu J, et al. Aβ-mediated NMDA receptor endocytosis in Alzheimer's disease involves ubiquitination of the tyrosine phosphatase STEP61. *J Neurosci* 2010;**30**(17):5948−57.

143. Yan SD, Chen X, Fu J, et al. RAGE and amyloid-β peptide neurotoxicity in Alzheimer's disease. *Nature* 1996;**382**(6593):685−91.

144. Chen X, Walker DG, Schmidt AM, Arancio O, Lue LF, Yan SD. Review. RAGE: a potential target for Aβ-mediated cellular perturbation in Alzheimer's disease. *Curr Mol Med* 2007;**7**(8):735−42.

145. Lue LF, Walker DG, Brachova L, Beach TG, Rogers J, Schmidt AM, et al. Involvement of microglial receptor for advanced glycation endproducts (RAGE) in Alzheimer's disease: identification of a cellular activation mechanism. *Exp Neurol* 2001;**171**:29.

146. Takuma K, Fang F, Zhang W, et al. RAGE-mediated signaling contributes to intra-neuronal transport of amyloid-β and neuronal dysfunction. *PNAS* 2010;**106** (47):20021−6.

147. Giri R, Shen Y, Stins M, Du Yan S, Schmidt AM, Stern D, et al. Beta-amyloid-induced migration of monocytes across human brain endothelial cells involves RAGE and PECAM-1. *Cell Physiol* 2000;**279**:C1772−81.

148. Dearie R, Sagare A, Zlokovic BV. The role of the cell surface LRP and soluble LRP in blood-brain barrier Aβ clearance in Alzheimer's disease. *Curr Pharm Des* 2008; **14**(16):1601−5.

149. de Chaves EP, Narayanaswami V. Review. Apolipoprotein E and cholesterol in aging and disease in the brain. *Future Lipidol* 2008;**3**(5):505−30.

150. Kim J, Basak JM, Holtzman DM. Review. The role of apolipoprotein E in Alzheimer's disease. *Neuron* 2009;**63**(3):287−303.

151. Bu G. Review. Apolipoprotein E and its receptors in Alzheimer's disease: pathways, pathogenesis and therapy. *Nat Rev Neurosci* 2009;**10**(5).

152. Bales KR, Verina T, Cummins DJ, et al. Apolipoprotein E is essential for amyloid deposition in the APP(V717F) transgenic mouse model of Alzheimer's disease. *PNAS* 1999;**96**(26):15233−8.

153. LaFerla FM, Troncoso JC, Strickland DK, Kawas CH, Jay G. Neuronal cell death in Alzheimer's disease correlates with apoE uptake and intracellular Aβ stabilization. *J Clin Invest* 1997;**100**(2):310−20.

154. Han SH, Hulette C, Saunders AM, et al. Apolipoprotein E is present in hippocampal neurons without neurofibrillary tangles in Alzheimer's disease and in age-matched controls. *Exp Neurol* 1994;**128**(1):13−26.

155. Han SH, Einstein G, Weisgraber KH, et al. Apolipoprotein E is localized to the cytoplasm of human cortical neurons: a light and electron microscopic study. *J Neuropathol Exp Neurol* 1994;**53**(5):535−44.

156. Bu G, Cam J, Zerbinatti C. Review. LRP in amyloid-β production and metabolism. *Ann N Y Acad Sci* 2006;**1086**:35−53.

157. Qiu Z, Strickland DK, Hyman BT, Rebeck GW. Alpha2-macroglobulin enhances the clearance of endogenous soluble β-amyloid peptide via low-density lipoprotein receptor-related protein in cortical neurons. *J Neurochem* 1999;**73**:1393−8.

158. Narita M, Holtzman DM, Schwartz AL, Bu G. Alpha-macroglobulin complexes with and mediates the endocytosis of β-amyloid peptide via cell surface low-density lipo-protein receptor-related protein. *J Neurochem* 1997;**69**(5):1904−11.

159. Wilhelmus MMM, Otte-Holler I, Davis J, Van Nostrand WE, de Waal RMW, Verbeek MM. Apolipoprotein E genotype regulates amyloid-β cytotoxicity. *J Neurosci* 2005;**25** (14):3621−7.

160. Van Uden E, Mallory M, Veinbergs I, Alford M, Rockenstein E, Masliah E. Increased extracellular amyloid deposition and neurodegeneration in human amyloid precur-sor protein transgenic mice deficient in receptor associated protein. *J Neurosci* 2002;**22** (21):9298−304.

161. Gylys KH, Fein JA, Tan AM, Cole GM. Apolipoprotein E enhances uptake of soluble but not aggregated amyloid-β protein into synaptic terminals. *J Neurochem* 2003; **84**(6):1442−51.

162. Yamada K, Hashimoto T, Yabuki C, et al. The low density lipoprotein receptor-related protein 1 mediates uptake of amyloid β peptides in an in vitro model of the blood-brain barrier cells. *J Biol Chem* 2008;**283**(50):34554−62.

163. Kang DE, Pietrzik CU, Baum L, et al. Modulation of amyloid β-protein clearance and Alzheimer's disease susceptibility by the LDL receptor-related protein pathway. *J Clin Invest* 2000;**106**(9):1159−66.

Natural Intracellular Consequences of Amyloid

Aβ, like its parent APP, is a normal, biologically multifunctioning peptide in spite of the widely publicized pathological role of Aβ (and its fragment, Aβ1-42 [Aβ42]) in AD. If Aβ is toxic, which is the basis of the amyloid cascade hypothesis, then a logical goal to treat AD would be to remove and inhibit the production of Aβ to reduce toxicity. This concept has been the objective in many clinical trials. As noted earlier in

Intracellular Consequences of Amyloid in Alzheimer's Disease.
DOI: http://dx.doi.org/10.1016/B978-0-12-804256-4.00003-6

this book, by simply reducing the levels of Aβ in the plasma, CSF and/or in the brain of patients with AD is not only ineffective but can also produce serious adverse effects. One very strong explanation is that Aβ is not toxic, and removing it from the body will have serious consequences. I commented on this very topic in a recent blog entitled "Are amyloid and the neuron innocent accomplices in AD?"[1] If indeed, amyloid, or more specifically, Aβ, is an innocent bystander in the pathological processes leading to AD, then what is its purpose in normal biological functions in the body?

Evolutionarily speaking, APP is an ancient, highly conserved protein that became increasing more complex during evolution; just this fact alone implies that APP is important to life (Fig. 3.1).[2–5] Note that APP alone does not always imply Aβ, or its derivative Aβ42, but it does imply that APP and its fragments do play a critical role in life in a plethora on normal biological processes involving neurons and other cell types. Some of the reported functions of Aβ in neurons include the processes of synaptic plasticity, learning and memory, and

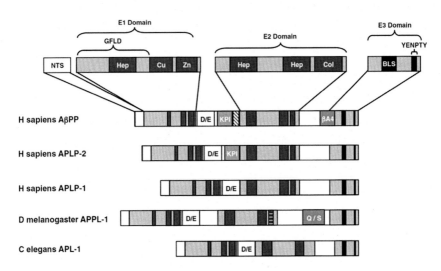

FIGURE 3.1   Conserved regions of the AβPP gene family. Schematic representations of the five members of the gene family show multiple conserved domains: N-terminal signal peptide (NTS), growth factor–like domain (GFLD), heparin-binding domain (Hep), copper-binding domain (Cu), zinc-binding domain (Zn), an acidic amino acid–rich region (D/E), collagen-binding domain (Col), a basolateral sorting signal (BLS), and a clathrin-binding internalization signal domain (YENPTY). Certain members of the gene family also contain a Kunitz-protease inhibitor (KPI) domain, the amyloid-β-forming region (βA4), an OX-2 domain (diagonal hashmarks), a putative collagen-binding domain (horizontal white hashmarks), and/or a glutamine- and serine-rich region (Q/S). *AβPP*, amyloid-β precursor protein; *APLP*, amyloid precursor-like protein; *APPL*, AβPP-like 1 protein; *APL*, amyloid precursor-like 1 protein. *Source: Used with permission from BMC Genom 2013;14:290.*

survival. In nonneuronal cells, Aβ plays a role in cellular proliferation, wound healing, and can act as an antimicrobial. Although Aβ may have a toxic role in the pathological processes leading to AD (see Chapter 4), it is clear from the reported science that Aβ is required for life.

Aβ FUNCTIONS IN NORMAL NEURONS

In 1987, Aβ was detected in normal human brains.[6−8] A few years later, Aβ42, the reported toxic Aβ species, was also detected in the neurons of normal brains.[9−13] Aβ is also expressed in the neurons of monkey brains as young as 4 years old.[14] Over time, the morphology of the intraneuronal Aβ changed from diffuse granules (4 years old) to clusters or clumps in the advanced age monkeys (24−36 years old), which was similarly observed in the APP transgenic mice with the Swedish and Arctic mutations.[15] This diffuse pattern of immunoreactivity indicated that the intracellular Aβ in young monkeys could represent Aβ generated naturally in neurons, rather than accumulated and internalized Aβ that is typically found in neurons of aged animals. This clustering of Aβ in neurons is consistent with previous hypotheses that Aβ accumulates in lipid rafts, and that clustering of GM1-type ganglioside within lipid raft-like membranes may facilitate the aggregation of Aβ (fibrils) in these rafts that are likely to correlate with increasing intramembrane cholesterol content during aging.[14] Although the biochemical analyses showed the presence of Aβ40 and Aβ43 in the microsomes of the young monkey brains, during aging, the generation of intraneuronal Aβ shifted to increased levels of Aβ42 while the Aβ40 remained unchanged, which may explain pathological events leading to AD (see Chapter 4).

Neuronal Development

Accumulating evidence suggests that Aβ may have an important physiological role in synapse elimination during brain development (Fig. 3.1).[16] It is important to note that the human Aβ revealed a high degree of homology (96.8%) to the mouse homologue. The calculated evolutionary rate of the mRNA at amino acid substitution site was relatively low $(0.1 \times 10^{-9}/\text{site}/\text{year})$, and coupled with the abundant presence in the brain, kidney, and in other tissues to a low level, it was hypothesized that "Aβ is highly conserved through mammalian evolution and may be involved in basic biological processes" such as neuronal development.[17]

To assess the function of APP, its expression was investigated in the rat brain in a time-course study from the embryonic to the postnatal stage of development.[18] The first expression of APP was detected at embryonic day (E15). Spinal anterior horn cells, spinal and trigeminal ganglion cells, trigeminal nerve nuclei are the first APP detectable neurons at E15. In addition to these neurons, ependymal cells, choroid plexus epithelial cells, and cells in the eye also revealed intense immunoreactivity at E15. By postnatal day 7 (P7), most neurons in the CNS and the peripheral nervous system expressed APP. The neurons at every site gradually gained the immunoreactivity of APP parallel to the age.

The membrane-spanning localization of APP suggests a role in cell−cell or cell−matrix interaction.[19,20] Immunoco-localization of APP and synaptophysin, a marker of synapses, confirmed the presence of APP at synaptic sites.[21] APP was also detected in the Purkinje cell body and dendrites of the normal cerebellum suggesting a role for APP in the growth of dendrites in Purkinje cells.[22] Increased synaptophysin expression coincided with APP expression in the cerebellar molecular layer and correlated with the timing of dendrite's outgrowth of Purkinje cell. These characteristic immunostaining patterns of APP were observed only after 2 weeks from P0. As for the age of P14, the dendrites APP staining of Purkinje cell became obscure because abundant neuropil staining of APP appeared in the molecular layer in a similar fashion to immunolabeling pattern of synaptophysin. These data suggested that APP plays an important role in neuronal maturation, and synaptogenesis follows APP expression.

Viability

About 90% of the total secreted Aβ is in the Aβ40 form, although Aβ42 is also normally produced and secreted by cells. However, both species are necessary for neuronal survival (eg, trophic and neuroprotective actions) at their normal physiological concentrations.[23−26]

Synthetic Aβ42 monomers (30−100 nanomolar (nM)) protect mature neurons against excitotoxic death of developing neurons under conditions of trophic deprivation.[27] In cultured neural stem cells, Aβ42 increased the number of newborn neurons, and it exhibited highly protective effects not only when combined with 100 nM NMDA toxicity, but also when applied before or after the NMDA pulse. In addition, pM levels of Aβ40 monomers were also fully protective against NMDA toxicity, and reversed the toxicity of secretase inhibition.[28] Inhibition of endogenous Aβ production by exposure to inhibitors either of β- or γ-secretases in primary neuronal cultures caused neuronal cell death.[29] These findings provide evidence for a role for Aβ in neuronal survival.[23]

Aβ also activates insulin receptors, which have neuroprotective effects. For example, when they are activated by insulin, they facilitate glucose utilization, energy metabolism, synaptic plasticity, and neuronal survival in the normal brain.[23] In fact, insulin has a positive effect on improved memory in rats and most importantly, in normal and AD adults.[30,31] Activation of the insulin receptors leads to activation of the phosphoinositide-3-kinase pathway (PI3K), which is a membrane-associated second messenger protein, and activated the downstream protein kinase B (Akt), which is also associated with neuronal survival and plasticity via activation of transcription pathways and protein synthesis.[23]

Although Aβ is not toxic to astrocytes, it does induce functional changes to these cells.[32] Aβ can alter the expression of various cytokines and other active molecules including interleukin-1β, interleukin-6, basic fibroblast growth factor, and nitric oxide synthase or promote glutamate clearance capacity of astrocytes by upregulating the glutamate asparate transporter.[33–35] In addition, Aβ can induce prominent increases in tyrosine phosphorylation of the extracellular signal-regulated kinase (ERK) 1 and 2 resulting in their translocation from the cytosol to the nucleus that was completely blocked by a specific inhibitor, thereby implying that they play a role in the regulation of gene transcription.[36] In neurons, ERK signaling pathways are also critical for memory and are tightly regulated as evident through human mental retardation syndromes.[37]

Arborization

The mechanisms regulating the outgrowth of neurites (ie, arborization) during development, as well as after injury, are key to the understanding of the wiring and functioning of the brain under normal and pathological conditions. The APP is also conserved in invertebrates and its *Drosophila* homologue, named APP-like (APPL), is also expressed in neurons.[38] APPL does not contain an Aβ peptide sequence, suggesting that the conserved physiological functions of APP do not involve Aβ. Although triple knockout mice for APP and its homologues are lethal and show focal cortical dysplasia, flies mutant for *appl* are viable, fertile, and show some neurological defects, such as disturbed synapse formation and axonal transport as well as developmental defects in the peripheral sensory organ formation.[39–41] Some studies have attributed a role to the APP extracellular domain in neurite outgrowth, neuronal survival, and proliferation, but it is the intracellular domain that is critical in the processes of postdevelopment axonal arborization, especially in the processes of neuronal repair.[42–44]

In a model of brain damage in the adult *Drosophila*, expression of the APPL is increased for several days, specifically in neurons surrounding damaged brain areas after injury, and flies mutant for *appl* show increased posttraumatic mortality suggesting that APPL has an important physiological role under these conditions. Many neurons in the injured brain areas also show activation of the Jun N-terminal kinases (JNK) signaling cascade, which play vital roles in many physiological and pathological processes. JNK signaling along with activation of the Abelson (Abl) tyrosine kinase are essential for correct regeneration of axons after injury in mammals.[44] APP-induced axonal arborization depends on intact JNK signaling. While expression of the membrane-bound C-terminal fragment of APP (APP-CTF) is sufficient to induce axonal arborization, expression of the AICD does not suggest that the cleavage of APP-CTF by γ-secretase may terminate APP activity and therefore regulate APP-Abl signaling.[44]

In this model, acute trauma resulted in increased APPL expression and independent, but simultaneous, JNK signaling activation. JNK activity provides, via transcriptional activation of profilin, the actin-binding protein that is required for the restructuring of actin filaments and other cytoskeletal regulators that are involved in remodeling of the actin cytoskeleton.[45]

Other family members of APP also promote neuritogenesis. APP-like proteins (APLP) 1 and 2 are also part of the APP gene family and the APLP2 has the greatest homology to APP and shares similar patterns of expression in the human and mouse.[46–48] When APLP2 −/− and APP −/− mice were crossed to generate double knockout mice, they showed a reduction in viability with only a 26% survival rate indicating the requirement of APP and APLP2 for postnatal survival.[49] Like APP, APLP2 can also promote neurite outgrowth in vitro in chick sympathetic neurons when exposed to purified recombinant APLP2 secreted from yeast *Pichia pastoris* transfected with the human APLP2 ectodomain.[50–55] The retarded neurite development of APP −/− hippocampal neurons and the finding that sAPLP2 has greater neurite length-promoting activity than secreted APP suggests APP and APLP2 have distinct roles in controlling neurite development.[55]

APP is also present in the growing tips of nerve fibers and is associated with the rapid elongation of axons.[56] Blocking APP expression with antisense oligonucleotides was enough to decrease axonal and dendritic outgrowth in embryonic cortical neurons in vitro, which again suggests important functions of APP in processes of neuronal outgrowth.[57]

Synaptic Physiology

Synaptic activity is required for neurons to survive through entry of appropriate amounts of calcium through synaptic receptors and channels.[58,59] Properly controlled homeostasis of calcium signaling not only supports normal brain physiology but also maintains neuronal integrity and long-term cell survival. Calcium signaling pathways can suppress apoptosis and promote survival through two mechanistically distinct processes. One process involves the PI3K/protein kinase B signaling pathway, which promotes survival.[60] The other pathway requires the generation of calcium transients in the cell nucleus, which offers long-lasting neuroprotection.[59,61]

At physiological concentrations, Aβ is essential for synaptic physiology, regulating synaptic scaling, synaptic release, and synaptic plasticity, which can be defined as the process by which neuronal synapses modulate their strength and form new connections with other neurons that play particularly important roles in learning and memory as well as response to injury and disease.[23,62−65]

Increases in Aβ-induced cytosolic calcium signals are also transmitted to protein kinase C (PKC) for PKC-mediated transcriptional activation.[23] PKC is a member of the protein kinase enzyme family that is responsible for controlling the functions of other proteins and also has a prominent role in memory.[66] Malfunctioning of calcium signaling to the cell nucleus can lead to neurodegeneration and neuronal cell death.[67]

Aβ is also a regulator of ion channel function and is essential for neuronal health.[29] For example, the α7 receptor modulates calcium homeostasis and release of the neurotransmitter acetylcholine, two important parameters involved in cognition and memory.[68] Receptor-binding experiments show that Aβ42 binds with high affinity to the α7 receptor and this interaction can be inhibited by α7 receptor ligands.[69,70] Aβ42 administered in the low pM range activates α7 receptors at presynaptic nerve endings of synaptosomes.[8,70] Under normal conditions, activation of α7 receptors is necessary for the Aβ-induced increase in synaptic plasticity and memory.[71] Aβ also directly interacts with calcium channels such as voltage-dependent calcium channels and TRP cation channels (TRPC) to produce a transient increase in calcium necessary for synaptic plasticity and neuronal survival. Aβ interacts directly with the recombinant L-type calcium channel α1C subunit to increase the calcium channel protein at the cell membrane and hence increase calcium conductance.[64] Within the TRPC subfamily, TRPC3 and -6 protect cerebellar granule neurons against serum deprivation−induced cell death in cultures and promote neuronal survival in rat brain.[72] A neuronal survival mechanism of Aβ can also involve altered expression of potassium channels.[64] In cerebellar granule neurons, 24-h preincubation with 1 micromolar (μM) unaggregated Aβ protein resulted in a 60% increase in the "A"-type component of potassium current possibly reflecting calcium-mediated gene expression.[73]

In vivo studies on wild-type animals and in vitro studies on wild-type and knockout animals demonstrated increased Aβ production with communication activity between brain cells that can subsequently downregulate synaptic transmission and neuronal activity.[62,71,74–76] This negative feedback loop could operate as a physiological homeostatic mechanism to limit levels of neuronal activity.[77] Neuronal activity and synaptic function can alter the levels of Aβ: blocking neuronal activity with tetrodotoxin treatment reduced Aβ levels, while increasing neuronal activity with picrotoxin enhanced Aβ secretion.[62] The processes regulating APP production and processing in response to neuronal activity could be attributed to γ-secretase function.[62,78]

Endogenous Aβ peptides also have a crucial role in activity-dependent regulation of synaptic vesicle release.[63] Presynaptic enhancement of the hippocampal neuronal network was mediated by Aβ and depended on the history of synaptic activation, with lower impact at higher firing rates. Changes in Aβ levels attenuated short-term synaptic facilitation during bursts in excitatory synaptic connections. Together, these observations support that Aβ secretion could be a physiologic event occurring with normal brain activity.

Learning and Memory

In contrast to the reported functional impairments caused by Aβ and its fragments in AD, Aβ peptides not only play a role in the cellular processes of learning and memory but can also enhance learning and memory.

Increased cognitive activity increases Aβ secretion since recovery of cognitive function after brain injury correlates with increased levels of extracellular Aβ in human brain.[79] Levels of Aβ in the interstitial fluid follow the circadian rhythm, being elevated when awake and reduced when asleep, which may align with cognitive activity as well.[80] Changes in the levels of Aβ can determine whether Aβ has a beneficial or toxic effect. For example, pM concentrations (200 pM) close to those in the normal brain of both Aβ42 monomers and oligomers cause a marked increase in LTP, whereas high nM concentrations (200 nM) lead to the well-established reduction of potentiation in the hippocampus.[81–85] Aβ secretion could be a physiologic event occurring with normal brain activity. With activity-induced secretion, the concentration of extracellular Aβ remains in the pM range in vitro as well as in vivo suggesting extracellular Aβ is nontoxic at low concentrations and play a novel positive modulatory role on neurotransmission, synaptic plasticity, and memory, whereas high concentrations are associated with neuronal cell death.[85–88]

Acute treatment of young (70–120 days) rats' hippocampal slices with low concentrations (100–200 nM) of bath-applied Aβ40 did not change basal synaptic transmission but showed an increase in tetanus-induced LTP.[89] Intracellular (100 nM via the recording pipette) or bath (200 nM) application of Aβ40 triggered the slow onset potentiation of NMDA receptor–mediated synaptic currents in hippocampal slices from young rats (70–120 g weight), and did not affect basal α-amino-3-hydroxyl-5-methyl-4-isoxazole-propionate (AMPA) receptor–mediated transmission, resting membrane potential, or input resistance of the granule cells.[90,91] Similar results showed no effect of Aβ42 on AMPA currents and demonstrated an increase in NMDA currents by the peptide.[92]

Data from the APP knockout mice and β-site APP-cleaving enzyme (BACE) 1 knockout mice showed impaired LTP as well as spatial and avoidance learning and memory.[93–95] *Drosophila* lacking the fly homologue for APP (*appl*) shows impaired avoidance learning that can be rescued by the expression of the human APP gene.[96] The use of an antibody or antisense-mediated to block APP during an early phase of memory formation also disrupted inhibitory avoidance in chicks.[97] Intracerebral or intraventricular administration of antibodies that bind to various domains of APP, including the middle portion of the Aβ fragment, disrupts learning and memory in rats, suggesting that the role of APP in memory formation could be mediated by Aβ.[98,99]

The effect on memory retention was assessed by either blocking endogenous Aβ or enhancing Aβ42 concentration in the hippocampus, a region known to be critically involved in the formation of explicit memories in a model of inhibitory avoidance.[100] This model is based on a fear-conditioning task where rats learn to associate context with an adverse stimulus (a mild foot shock), and subsequently learn avoidance, which is a hippocampus-based neuronal process.[88] Blocking in vivo Aβ function with a monoclonal antibody (mAb) completely disrupted short-term memory at 1 h after training, as well as long-term memory 24 h after training, which persisted for 5 days after training that was not observed in locomotor and nociceptive tests, whereas no effect was observed when the anti-Aβ mAb was injected after training (Fig. 3.2).

Remarkably, memory was rescued when bilateral injections of exogenous human Aβ42 (100 pM) were given to those rats also receiving the anti-Aβ mAb. When this exogenous human Aβ42 was given to those rats that received the placebo injection of a control mAb, they showed significantly enhanced memory retention.[88] As the control experiment, rats that received the initial injection of the anti-Aβ mAb and then were injected with the scrambled control peptide, sustained memory impairment. This was also observed with intraventricular administration of the secreted form of APP that produced memory enhancement

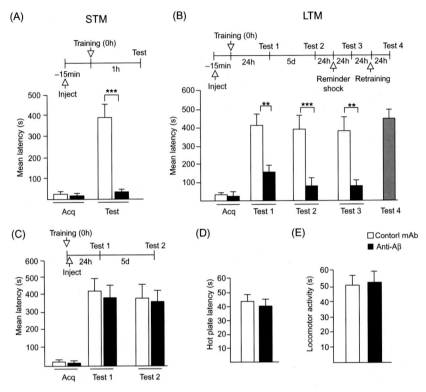

FIGURE 3.2   Depletion of endogenous Aβ disrupts memory retention. Memory acquisition and retention are expressed as mean latency ± SEM (in seconds, s). Rats received intrahippocampal injections of either anti-Aβ or control mAb antibody 15 min before IA training and tested for STM (A) or LTM (B) at 1 or 24 h after training, respectively. Both STM and LTM were disrupted by the anti-Aβ antibody. LTM disruption persisted 5 days after training, and memory did not recover following a reminder foot shock administered in a different context a day later. Amnesic rats that received the anti-Aβ after retraining showed normal retention. ** $P < 0.01$, *** $P < 0.001$. (C) Intrahippocampal injections of either anti-Aβ or control mAb immediately after training had no effect on memory tested at 24 h and 5 days after training. (D) No effect of intrahippocampal injections of anti-Aβ on the nociceptive hot plate test. Rats injected with anti-Aβ or control mAb underwent the hot plate test 15 min after injection. Values are expressed as the mean ± SEM of response latencies measured in seconds. (E) No effect of intrahippocampal injection of anti-Aβ antibody on locomotor activity. One hour after injection of either anti-Aβ or control mAb antibody, rats were allowed to explore the IA training apparatus for 3 min and the locomotor activity was detected by a system of photocell infrared beams. Values are expressed as mean ± SEM of motility counts. *IA,* inhibitory avoidance; *LTM,* long-term memory; *mAb,* monoclonal antibody; memory acquisition (acq); *SEM,* standard error mean; STM: short-term memory. *Source: Used with permission from* Learning & Memory 2009;**16**:267−72.

that rescued the amnesia induced by the cholinergic antagonist scopol-amine.[101] These results indicated that endogenous Aβ plays a critical role during learning for both short- and long-term memory formation, and that Aβ42 in low concentrations (pM range) can enhance memory function.

Since Aβ42 binds selectively and with pM affinity to the α7 receptors, the experiments were repeated only this time the exogenous Aβ42 was replaced with mecamylamine, a cholinergic receptor antagonist, either 15 min before or 1 h after training.[69] The same results were obtained; blocking the α7 receptors also produced memory impairments that did not affect locomotor activity or nociceptive sensitivity. Similar behavioral outcomes were produced by hippocampal treatments that block the function of either endogenous Aβ or nicotinic acetylcholine receptors suggesting that Aβ42 might exert its modulatory function on memory formation via interaction with these receptors.[11,69,102–106]

Low doses of Aβ enhanced memory retention in two memory tasks and enhanced acetylcholine production in the hippocampus in vivo.[106] To further test this possibility, endogenous Aβ was blocked in young, healthy, cognitively intact mice. The methods of blocking endogenous Aβ were (1) with antibody to Aβ, or (2) DFFVG (which blocks Aβ binding), or (3) decreasing Aβ expression with antisense directed at the Aβ precursor, APP. The results showed that all methods blocking endogenous Aβ resulted in impaired learning in T-maze foot-shock avoidance. Finally, Aβ42 facilitated the induction and maintenance of LTP in hippocampal slices, whereas antibodies to Aβ inhibited hippo-campal LTP. These results further indicated the importance of Aβ in learning and memory.[106]

BACE1 −/− Mice

BACE1 has normal physiological functions such as in the processes of synaptic transmission and plasticity.[107–109] In organotypic hippocampal slices, BACE activity is increased by synaptic activity, and the resulting Aβ peptides depress excitatory transmission through AMPA and NMDA receptors, suggesting a role for Aβ in homeostatic plasticity.[58] Because BACE1 is initially responsible for producing the reported toxic Aβ species (Aβ40 and Aβ42) that contribute to the AD pathology, a mouse knockout model of AD was created. Without BACE1, there was no β-secretase activity and essentially no Aβ (Aβ40 and Aβ42) production as compared to littermates.[107]

These BACE1 knockout mice (BACE −/−) were viable and fertile with no gross differences in behavior or development.[107–109] However, upon further analysis, these mice presented impaired synaptic plasticity, memory, and cognition suggesting a necessary role of Aβ.[108] Specifically, young BACE1 −/− mice (3−6 months old) displayed specific deficits in paired-pulse facilitation and de-depression implicating significant alterations in mechanisms of presynaptic release and synaptic plasticity. Severe deficits in presynaptic function were also observed at the synapses including a reduction in presynaptic release.

The strengthening of synaptic connections, also referred to as LTP, is experimentally produced by high-frequency stimulation, while the weakening of synaptic connections produced by low-frequency stimulation is called long-term depression (LTD).[107] The BACE1 −/− mice exhibited a slightly larger mossy fiber LTD, which could not be reversed suggesting that BACE1 function is crucial for normal synaptic transmission and activity-dependent presynaptic potentiation at these synapses. Although the majority of studies characterizing synaptic function of BACE1 −/− mice were performed in the CA1 region of the hippocampus, the expression of BACE1 is most prominent in the mossy fiber terminals that synapse onto CA3 pyramidal neurons. The BACE1 −/− mice also displayed severe deficits in presynaptic function at these synapses, including a reduction in presynaptic release and an absence of mossy fiber LTP, which is normally expressed by a long-term increase in presynaptic release.

The auxiliary β2 subunit of the voltage-gated sodium channel (Nav1), a critical channel for generating action potentials, is also a substrate of BACE1.[107] BACE1 regulates the surface expression of the four different CNS types of Nav1 channels by cleaving the β2 subunit. BACE1 −/− mice have a decrease in Nav1 expression (mRNA and protein) suggesting a role for BACE1 in regulating neuronal excitability.

BACE1 is also important for myelination of axons.[107,109] Neuregulin 1, an axonal signaling molecule critical for regulating myelination, is another substrate of BACE1. BACE1 −/− mice show hypomyelination in the central nerves. Neuregulin 1 and its receptor, ErbB4, regulate synaptic function and plasticity by affecting presynaptic release by regulating α7 receptors. These results further strengthen the necessary function of BACE1 in pre- and postsynaptic function.

As noted, the deletion of BACE1 in mice did not lead to overt developmental abnormalities but additional behavioral studies showed mild cognitive deficits. The Morris water maze task was developed to assess spatial reference memory through the need to learn and remember the hidden platform location. The BACE1 −/− mice exhibited age-dependent deficits in this maze, meaning that, over time, they were able to adopt the needs of the task. However, in the radial water maze, the BACE −/− mice were significantly impaired compared to the BACE +/− and BACE +/+ mice. In the Y-maze task, an independent method to assess spatial working memory, the BACE −/− mice had deficits similar to the radial water maze suggesting that loss of BACE1 resulted in the early cognitive deficits in spatial working memory.[110] Taken together, these findings support the view that BACE1 plays a critical role in both spatial reference memory and working memory.[108]

Aβ FUNCTIONS IN OTHER CELL TYPES

Aβ is also present on many other tissues besides the brain such as in the epidermis, thyroid, and pancreas.[13,15,111–118] The data presented in this section note that Aβ, its fragments, and APP not only have a ubiquitous distribution in the body but is also critically involved in many physiological processes outside of the brain.

For example, the epithelial cells of the thyroid produce large amounts of APP and generate potentially amyloidogenic APP fragments.[115] After analyzing the forms of APP detected in cultured thyroid epithelial cells, it was proposed that Aβ may be released into the circulation and may explain the frequent occurrence of thyroid autoimmunity in AD.

Fat Metabolism

APP and Aβ are present in epithelial cells of the small intestine.[116,117] Several lines of evidence suggest that Aβ is involved in the metabolism of dietary fats, and that aberrations in postprandial lipemia might be a contributing factor for AD. There was a transient increase in the plasma concentration of APP following the ingestion of dietary fats.[119] Epidemiological studies reported a positive correlation of fat intake with AD prevalence.[120,121] In animal studies, cerebral amyloidosis induced by high-fat feeding corresponded with dietary-induced hyperlipidemia and raised chylomicron concentration.[122–124] Therefore, diet can affect Aβ production in intestinal epithelial cells.

Prominent, apical intracellular Aβ immunoreactivity was detected in the columnar absorptive epithelial cells in the wild-type mice (Fig. 3.3). The presence of Aβ/APP within the perinuclear region demonstrates intracellular synthesis of Aβ from the rough ER and the Golgi apparatus, sites of chylomicron lipid pools.[116] The qualitative patterns of Aβ/APP distribution were similar for mice fed the high-fat and low-fat diets. However, significantly greater intestinal epithelial expression of Aβ/APP in mice given the high-fat diet ($P < 0.001$) was detected (Fig. 3.3B). Ingestion of saturated fat significantly enhanced enterocytic Aβ abundance whereas fasting abolishes expression.[117] Coupled with the findings that Aβ binds avidly with chylomicrons, the data suggest

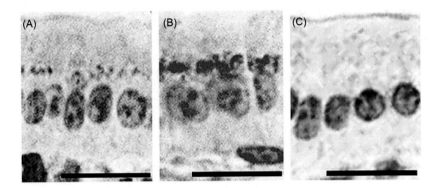

FIGURE 3.3   Enterocytic expression of Aβ/APP in 6-month-old WT mice (C57BL/6J) fed a low-fat or a high-fat diet for 6 months or in mice fasted for 65 h. High magnifications of Aβ/APP immunostaining in enterocytes of the small intestine of mice fed a low-fat or a high-fat diet are shown in (A) and (B), respectively. Substantial staining for Aβ/APP was detected within the perinuclear region of the cytoplasm, which was amplified by high-fat feeding. Note the increase in the size and density of Aβ/APP granules in absorptive cells in mice subjected to 6 months of high-fat diet (B) compared to low-fat diet (A). (C) Aβ/APP in epithelial cells of WT mice maintained on a high-fat diet but deprived of food for 65 h prior to sacrifice. Scale bar = 5 μm. *Aβ*, amyloid-β; *APP*, amyloid precursor protein; *WT*, wild-type. *Source: Used with permission from* J Nutr Biochem 2007;*18:279–84.*

that chronic high-fat feeding can increase intestinal Aβ association with chylomicrons, and that Aβ can regulate dietary fat metabolism.

Mitogenic, Cell Growth, and Migration Activity

Both the NH2- and COOH-terminal domains of APP mediate various functions including cell growth and process extension, and act as a mitogen.[13,118] High expression of APP was detected in pancreatic adenocarcinoma cancer cells using RT-PCR, western blotting, and immunocytochemistry.[13] IHC analysis of pancreatic adenocarcinoma tissues showed APP expression in pancreatic acinar and islet cells, but not in the ductal epithelial cells. Since the APP-expressing cancer cells originate from the APP-negative normal epithelial cells, the upregulation of APP expression suggests a role in pancreatic carcinogenesis (eg, proliferation, cell growth, and/or migration).

Secreted APP was necessary for epidermal growth factor–induced proliferation of progenitor cells in the adult subventricular zone suggesting an endogenous function of sAPP in adult neurogenesis.[125] APP also enhanced the proliferation of neural stem cells derived from fetal rat brain.[55] Specifically, both secretory APPs (sAPP770 and sAPP695, with and without the Kunitz-type serine protease inhibitor domain, respectively) promoted the growth of neural stem cells, and the effect of sAPP770 was greater than that of sAPP695. The sAPP770 peptide in combination with growth factors exerted a cooperative stimulation of the stem cell proliferation. These results suggest that APP, especially APP possessing the protease inhibitor domain, can regulate the growth of neuronal precursor cells during the development of the nervous system.

Incubation of fibroblasts and thyrocytes with sAPP also resulted in an upregulation of cell proliferation through the phosphorylation of mitogen-activated kinase in FRTL-5 thyroid epithelium cells, or by additional mechanisms such as PKC activation, Akt phosphorylation, or activation or potentiation of growth factor signaling pathways.[126–129]

Aβ oligomers can also promote neurogenesis, both in vitro and in vivo, by inducing neural progenitor cells to differentiate into neurons.[130,131] Treatment with oligomeric Aβ increased the proliferation and differentiation of neuronal progenitor cells in cultured choroid plexus epithelial cells, but decreased the survival of newly born neurons.[132] These Aβ-induced neurogenic effects were also observed in choroid plexus of APP/PS1 mice. Analysis of signaling pathways revealed that pretreating the choroid plexus epithelial cells with specific inhibitors of tyrosine kinase or mitogen-activated protein kinase reduced the Aβ-induced neuronal proliferation, thereby supporting a role of Aβ in the proliferation and differentiation in the choroid plexus epithelial cells.[132]

In addition to effects on the cell cycle, sAPP can alter cell morphology by inducing the extension of neurite processes in neuroblastoma cells and may also function as a growth factor, and its intracellular domain is thought to regulate transcription.[43,55,133–135]

Antimicrobial Activity

In the innate immune system, Aβ can also function as an antimicrobial peptide.[136] In particular, Aβ42 inhibited the overnight growth of clinically important pathogens representing Gram-positive, Gram-negative, and the yeast *Candida albicans* (Table 3.1). The synthetic Aβ42 peptide as well as Aβ42 derived from homogenized Alzheimer's brain tissues had potency equivalent to or greater than the synthetic LL-37, an archetypical human antimicrobial peptide, and much more activity than the Aβ40 peptide (synthetic or derived from the human tissues). These data suggest that removing Aβ can increase the risk of infection and can explain why patients who received taranflurbil, a γ-secretase modulator directed to lowering Aβ levels, had significantly increased rates of infection.[137,138]

Furthermore, patients deficient in LL-37 cannot mount an effective defense against pathogens.[139] Genetically manipulated mice to produce

TABLE 3.1 Aβ Peptides Possess Antimicrobial Activity

Organism	MIC (µg/mL)					
	Aβ42	Aβ40	roAβ42	LL-37	reAβ42	scAβ42
Candida albicans	0.78	0.78	0.78	6.25	>25	>50
Escherichia coli	1.56	1.56	3.13	1.56	>50	>50
Staphylococcus epidermidis	3.13	50	3.13	25	>50	>50
Streptococcus pneumoniae	6.25	12.5	6.25	1.56	50	>50
Staphylococcus aureus	6.25	25	12.5	6.25	>50	>50
Listeria monocytogenes	6.25	25	6.25	25	>50	50
Enterococcus faecalis	6.25	50	3.13	6.25	50	>50
Streptococcus agalactiae	12.5	50	>50	12.5	>50	>50
Pseudomonas aeruginosa	>50	>50	>50	6.25	>50	>50
Streptococcus pyogenes	>50	>50	>50	6.25	>50	>50
Streptococcus mitis	>50	50	>50	6.25	>50	>50
Streptococcus salivarius	>50	>50	>50	50	>50	>50

The antimicrobial activity of synthetic Aβ1-42 (**Aβ42**), Aβ1-40 (**Aβ40**), LL-37 (**LL-37**), reverse Aβ42-1 (**rAβ42**), or scrambled Aβ42 (**scAβ42**) peptides were determined as **MIC** against 12 microorganisms. Antimicrobial activity was assayed by broth microdilution susceptibility test on 96-well plates with microbial growth in wells determined by visual inspection following an overnight incubation. Inhibition of growth in plate wells was confirmed by alamar blue cell viability assay and by surface plating of incubants on agar and counting CFU. Inoculums contained mid-logarithmic phase cells. Consistent with antimicrobial activity specific to the Aβ sequence, inhibition was not observed for reverse and scrambled peptides. Aβ, amyloid-β; CFU, colony-forming units; MIC, minimal inhibitory concentrations.
Source: Used with permission from PLoS One 2010:5(3):e9505.

little to none Aβ (eg, BACE1 and BACE2 knockouts) have >50% survival rates that can be dramatically improved to survival levels (>95%) in the wild-type mice when housing these knockout mice in pathogen-free environments.[140]

Intracellular Cargo Receptor

In addition to mammals, APP is highly conserved in the *Xenopus*.[141] Both the Aβ and sAPPα are present in its pituitary melanotrope cells. The expression of APP was studied in these cells because the activities of these cells are easily manipulated by simply placing the frog on a black background. When adapting to a black background, the intermediate pituitary melanotrope cells upregulate the production of the prohormone pro-opiomelanocortin, the hormone responsible for skin darkening.[142] Conversely, when placing the *Xenopus* on a white background, the anterior melanotrope cells become inactive via their regulation by inhibitory neurons of hypothalamic origin.[143]

After placing the *Xenopus* on a black background, APP mRNA and protein were upregulated in the active intermediate pituitary melanotrope cells, whereas no changes in APP expression levels were detected in the anterior pituitary cells. This cell-specific induction of APP expression occurred because the melanotrope cells are controlled by neurons of hypo-thalamic origin that either activate or inactivate the cells depending on the color of the background of the animal, while the anterior pituitary cells are not involved in background adaptation.[143] Thus, APP may also be part of the biosynthetic machinery in the melanotrope cells. APP could also function in post-Golgi vesicular cargo transport as a membrane receptor for the microtubule-dependent motor protein kinesin-1, which was previously observed in axonal transport in neurons for the intact proteins.[144,145]

Wound Healing

APP and Aβ are expressed in many cell types in the skin including the keratinocytes, nerve and sensory corpuscles, and endothelial and smooth muscle vascular cells of all calibers.[12,146] Specific to the epidermis, APP is predominantly expressed in the keratinocytes of the basal cell layer of the normal skin.[146] During wound healing, APP is upregulated (protein and mRNA) in all layers of the hyperprolifera-tive epithelium at the wound margin. In vitro studies demonstrated significantly higher release of sAPP in proliferating keratinocytes than quiescent, partially differentiated keratinocytes. These data suggest that proliferating keratinocytes secrete APP in the margins of the wound to help promote growth in the processes of epidermal wound repair.

Also, the release of platelet Aβ by physiologic stimuli suggested that it may also play a role in platelet aggregation and coagulation as a repair mechanism associated with injuries like wound healing.[147]

SUMMARY

APP is an extremely complex and multifunctional molecule. Fig. 3.4 provides a summary of most of the reported functions of APP and its cleavage products.[148]

There is little doubt that APP and its derivatives (eg, Aβ and Aβ42) have essential roles in the regulation of key neural functions, including, but not limited to, synapse formation, maintenance, and growth, in neurite extension and in synaptic plasticity, and learning and memory.[149,150]

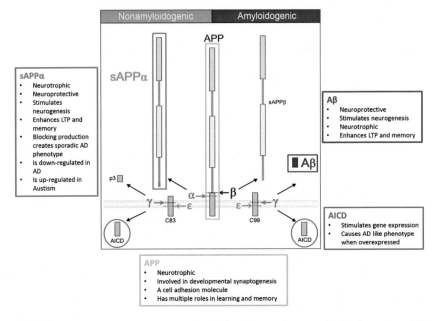

FIGURE 3.4   Cleavage of APP and physiological roles of APP and APP fragments. APP can be cleaved via two mutually exclusive pathways. In the so-called amyloidogenic pathway, APP is cleaved by BACE1 and γ-secretase enzymes (presenilin I is the catalytic core of the multiprotein γ-secretase complex). The initial β-secretase cleavage produces a large soluble extracellular domain, sAPPβ. The remaining membrane-bound C99 stud is then cleaved by multiple sequential γ-secretase cleavages. These begin near the inner membrane at a γ-secretase cleavage site ε (the ε-site) to produce the AICD, and then subsequent sequential γ-secretase cleavages trim the remaining membrane-bound component to produce different length Aβ peptides including Aβ43, Aβ42, Aβ40, and Aβ3. In the so-called nonamyloidogenic pathway, APP is processed consecutively by α- and γ-secretases to produce sAPPα, p3 (which is in effect Aβ17-40/42), and AICD. The major α-secretase enzyme is ADAM10. Cleavage via amyloidogenic and nonamyloidogenic pathways depends on the cellular localization of cleavage enzymes, and of full-length APP, which are expressed and trafficked in specific subcellular locations. *APP*, amyloid precursor protein; *AICD*, APP intracellular domain; *BACE1*, β-site APP-cleaving enzyme; *sAPPβ*, secreted amyloid precursor protein-β; *ADAM10*, A disintegrin and metalloproteinase domain-containing protein 10. *Source: Used with permission from Acta Neuropathol Commun 2014;2:135.*

Although amyloid, or specifically Aβ, is associated with AD, one has to wonder if Aβ is an innocent bystander being in the wrong place at the wrong time in the pathogenesis of the disease. The normal range of the Aβ peptides in the brain has been estimated to be about 200–1000 pM, with Aβ42 at the lower end of the range and Aβ40 at the higher end.[86,151] Perhaps when concentrations in the brain exceed these values is when Aβ becomes toxic as suggested by many studies. However, based on its myriad of essential biological functions, it would not be prudent to remove it all.

References

1. <http://scitechconnect.elsevier.com/amyloid-neuron-alzheimers/> [accessed 04.09.15].
2. Collin RWJ, Martens GJM. The amyloid-β precursor-like protein APLP2 and its relative APP are differentially regulated during neuroendocrine cell activation. *Mol Cell Neurosci* 2005;**30**:429−36.
3. Coulsona EJ, Paligab K, Beyreutherb K, Master CL. Review. What the evolution of the amyloid protein precursor supergene family tells us about its function. *Neurochem Int* 2000;**36**:175−84.
4. Tharp WG, Sarkar IN. Review. Origins of amyloid-β. *BMC Genomics* 2013;**14**:290−305.
5. Goldgaber D, Lerman MI, McBride OW, Saffiotti U, Gajdusek DC. Characterization and chromosomal localization of a cDNA encoding brain amyloid of Alzheimer's disease. *Science* 1987;**235**(4791):877−80.
6. Bahmanyar S, Higgins GA, Goldgaber D, Lewis DA, Morrison JH, Wilson MC, et al. Localization of amyloid beta protein messenger RNA in brains from patients with AD. *Science* 1987;**237**(4810):77−80.
7. Tanzi RE, Gusella JF, Watkins PC, Bruns GA, St George-Hyslop P, Van Keuren ML, et al. Amyloid beta protein gene: cDNA, mRNA distribution, and genetic linkage near the Alzheimer locus. *Science* 1987;**235**(4791):880−4.
8. Goedert M. Neuronal localization of amyloid beta protein precursor mRNA in normal human brain and in AD. *EMBO J* 1987;**6**(12):3627−32.
9. Gouras GK, Tsai J, Näslund J, Vincent B, Edgar M, Checler F, et al. Intraneuronal Aβ42 accumulation in human brain. *Am J Path* 2000;**156**(1):15−20.
10. D'Andrea MR, Nagele RG, Wang H-Y, Peterson PA, Lee DHS. Evidence that neurones accumulating amyloid can undergo lysis to form amyloid plaques in AD. *Histopathology* 2001;**38**:120−34.
11. D'Andrea MR. *Bursting neurons and fading memories*. New York: Elsevier Press; 2014
12. Vega JA, Diaz-Trelles R, Haro JJ, del Valle ME, Naves EJ, Fernandez-Sanchez MT. Beta-amyloid precursor protein in human digital skin. *Neurosci Lett* 1995;**192**:132−6.
13. Hansel DE, Rahman A, Wehner S, Herzog V, Yeo CJ, Maitra A. Increased expression and processing of the Alzheimer amyloid precursor protein in pancreatic cancer may influence cellular proliferation. *Cancer Res* 2003;**63**:7032−7.
14. Kakio A, Nishimoto SI, Yanagisawa K, Kozutsumi Y, Matsuzaka K. Cholesterol-dependent formation of GM1 ganglioside-bound amyloid-protein, an endogenous seed for Alzheimer amyloid. *J Biol Chem* 2001;**276**:24985−90.
15. Lord A, Kalimo H, Eckmang C, Zhang X-Q, Lars Lannfelt L, Nilsson LNG. The Arctic Alzheimer mutation facilitates early intraneuronal Aβ aggregation and senile plaque formation in transgenic mice. *Neurobiol Aging* 2006;**27**:67−77.
16. Wasling P, Daborg J, Riebe I, Andersson M, Portelius E, Blennow K, et al. Review. Synaptic retrogenesis and amyloid-β in Alzheimer's disease. *J Alzheimers Dis* 2009;**16**(1):1−14.
17. Yamada T, Sasaki H, Furuya H, Miyata T, Goto I, Sakaki Y. Complementary DNA for the mouse homologue of the human amyloid β protein precursor. *Biochem Biophys Res Commun* 1987;**149**(2):665−71.
18. Ohta M, Kitamoto T, Iwaki T, Ohgami T, Fukui M, Tateishi J. Immunohistochemical distribution of amyloid precursor proteinduring normal rat development. *Dev Brain Res* 1993;**75**:151−61.
19. Schubert D, Jin L-W, Saitoh T, Cole G. The regulation of amyloid β protein precursor secretion and its modulatory role in cell adhesion. *Neuron* 1989;**3**:689−94.
20. Shivers BD, Hilbich C, Mutthaup G, Salbaum M, Beyreuther K, Seeburg PH. Alzheimer's disease amyloidogenic glycoprotein: expression pattern in rat brain suggests a role in cell contact. *EMBO J* 1988;**7**:1365−70.

21. Schubert W, Prior R, Weidemann A, Dircksen H, Multhaup G, Masters CL, et al. Localization of Alzheimer β/A4 amyloid precursor protein at central and peripheral synaptic sites. *Brain Res* 1991;**563**:184−94.
22. Kawarabayashi T, Shoji M, Harigaya Y, Yamaguchi H, Hirai S. Amyloid β/A4 protein precursor is widely distributed in both the central and peripheral nervous systems of the mouse. *Brain Res* 1991;**552**:1−7.
23. Parihar MS, Brewer GJ. Review. Amyloid beta as a modulator of synaptic plasticity. *J Alzheimers Dis* 2010;**22**(3):741−63.
24. Dahlgren KN, Manelli AM, Stine WB, Baker LK, Krafft GA, LaDu MJ. Oligomeric and fibrillar species of amyloid-β peptides differentially affect neuronal viability. *J Biol Chem* 2002;**35**:32046−53.
25. Whitson JS, Selkoe DJ, Cotman CW. Amyloid beta protein enhances the survival of hippocampal neurons in vitro. *Science* 1989;**243**:1488−90.
26. Zou K, Gong JS, Yanagisawa K, Michikawa M. A novel function of monomeric amyloid beta-protein serving as an antioxidant molecule against metal-induced oxidative damage. *J Neurosci* 2002;**22**:4833−41.
27. Giuffrida ML, Caraci F, Pignataro B, Cataldo S, De Bona P, Bruno V, et al. Beta-amyloid monomers are neuroprotective. *J Neurosci* 2009;**29**:10582−7.
28. Yankner BA, Duffy LK, Kirschner DA. Neurotrophic and neurotoxic effects of amyloid beta protein: reversal by tachykinin neuropeptides. *Science* 1990;**250**:279−82.
29. Plant LD, Boyle JP, Smith IF, Peers C, Pearson HA. The production of amyloid beta peptide is a critical requirement for the viability of central neurons. *J Neurosci* 2003;**23**:5531−5.
30. Park CR, Seeley RJ, Craft S, Woods SC. Intracerebroventricular insulin enhances memory in a passive-avoidance task. *Physiol Behav* 2000;**68**:509−14.
31. Craft S, Watson GS. Insulin and neurodegenerative disease: shared and specific mechanisms. *Lancet Neurol* 2004;**3**:169−78.
32. Abe K, Hisatomi R, Misawa M. Amyloid β peptide specifically promotes phosphorylation and nuclear translocation of the extracellular signal-regulated kinase in cultured rat cortical astrocytes. *J Pharmacol Sci* 2003;**93**:272−8.
33. Pike CJ, Cummings BJ, Monzavi R, Cotman CW. Beta-amyloid-induced changes in cultured astrocytes parallel reactive astrocytosis associated with senile plaques in Alzheimer's disease. *Neuroscience* 1994;**63**:517−31.
34. Hu J, Akama KT, Kraft GA, Chromy BA, Van Eldik LJ. Amyloid-β peptide activates cultured astrocytes: morphological alterations, cytokine induction and nitric oxide release. *Brain Res* 1998;**785**:195−206.
35. Abe K, Misawa M. Amyloid β protein enhances the clearance of extracellular L-glutamate by cultured rat cortical astrocytes. *Neurosci Res* 2003;**45**:25−31.
36. Sugden PH, Clerk A. Regulation of the ERK subgroup of MAP kinase cascades through G protein-coupled receptors. *Cell Signal* 1997;**9**:337−51.
37. Sweatt JD. Review. Mitogen-activated protein kinases in synaptic plasticity and memory. *Curr Opin Neurobiol* 2004;**14**:311−17.
38. Martin-Morris LE, White K. The *Drosophila* transcript encoded by the β-amyloid protein precursor-like gene is restricted to the nervous system. *Development* 1990;**110**:185−95.
39. Herms J, Anliker B, Heber S, Ring S, Fuhrmann M, Kretzschmar H, et al. Cortical dysplasia resembling human type 2 lissencephaly in mice lacking all three APP family members. *EMBO J* 2004;**23**:4106−15.
40. Torroja L, Packard M, Gorczyca M, White K, Budnik V. The *Drosophila* β-amyloid precursor protein homolog promotes synapse differentiation at the neuromuscular junction. *J Neurosci* 1999;**19**:7793−803.
41. Gunawardena S, Goldstein LS. Disruption of axonal transport and neuronal viability by amyloid precursor protein mutations in *Drosophila*. *Neuron* 2001;**32**:389−401.

42. Merdes G, Soba P, Loewer A, Bilic MV, Beyreuther K, Paro R. Interference of human and *Drosophila* APP and APP-like proteins with PNS development in *Drosophila*. *EMBO J* 2004;**23**:4082–95.

43. Jin LW, Ninomiya H, Roch JM, Schubert D, Masliah E, Otero DA, et al. Peptides containing the RERMS sequence of amyloid β/A4 protein precursor bind cell surface and promote neurite extension. *J Neurosci* 1994;**14**:5461–70.

44. Leyssen M, Ayaz D, Hebert SS, Reeve S, De Strooper B, Hassan BA. Amyloid precursor protein promotes post-developmental neurite arborization in the *Drosophila* brain. *EMBO J* 2005;**24**:2944–55.

45. Wills Z, Marr L, Zinn K, Goodman CS, Van Vactor D. Profilin and the Abl tyrosine kinase are required for motor axon outgrowth in the *Drosophila* embryo. *Neuron* 1999;**22**:291–9.

46. Wasco W, Bupp K, Magendantz M, Gusella JF, Tanzi RE, Solomon F. Identification of a mouse brain cDNA that encodes a protein related to the Alzheimer disease-associated amyloid beta protein precursor. *PNAS* 1992;**89**:10758–62.

47. Slunt HH, Thinakaran G, Von Koch C, Lo ACY, Tanzi R, Sisodia SS. Expression of a ubiquitous, cross-reactive homologue of the mouse β-amyloid precursor protein (APP). *J Biol Chem* 1994;**269**:2637–44.

48. Webster MT, Groome N, Francis PT, Pearce BR, Sherriff FE, Thinakaran G, et al. A novel protein, amyloid precursor-like protein 2, is present in human brain, cerebrospinal fluid and conditioned media. *Biochem J* 1995;**310**:95–9.

49. von Koch CS, Zheng H, Chen H, Trumbauer M, Thinakaran G, Van der Ploeg LHT, et al. Generation of APLP2 KO mice and early postnatal lethality in APLP2/APP double KO mice. *Neurobiol Aging* 1997;**18**:661–9.

50. Clarris HJ, Nurcombe V, Small DH, Beyreuther K, Masters CL. Secretion of nerve growth factor from septum stimulates neurite outgrowth and release of the amyloid protein precursor of Alzheimer's disease from hippocampal explants. *J Neurosci Res* 1994;**38**:248–58.

51. Ninomiya H, Roch JM, Jin LW, Saitoh T. Secreted form of amyloid beta/A4 protein precursor (APP) binds to two distinct APP binding sites on rat B103 neuron-like cells through two different domains, but only one site is involved in neuritotropic activity. *J Neurochem* 1994;**63**:495–500.

52. Schubert D, LaCorbiere M, Saitoh T, Cole G. Characterization of an amyloid beta precursor protein that binds heparin and contains tyrosine sulfate. *PNAS* 1989;**86**:2066–9.

53. Small DH, Nurcombe V, Moir R, Michaelson S, Monard D, Beyreuther K, et al. Association and release of the amyloid protein precursor of Alzheimer's disease from chick brain extracellular matrix. *J Neurosci* 1992;**12**:4143–50.

54. Small DH, Nurcombe V, Reed G, Clarris H, Moir R, Beyreuther K, et al. A heparin-binding domain in the amyloid protein precursor of Alzheimer's disease is involved in the regulation of neurite outgrowth. *J Neurosci* 1994;**14**:2117–27.

55. Hayashi Y, Kashiwagi K, Ohta J, Nakajima M, Kawashima T, Yoshikawa K. Alzheimer amyloid protein precursor enhances proliferation of neural stem cells from fetal rat brain. *Biochem Biophys Res Commun* 1994;**205**(1):936–43.

56. Moya KL, Benowitz LI, Schneider GE, Allinquant B. The amyloid precursor protein is developmentally regulated and correlated with synaptogenesis. *Dev Biol* 1994;**161**(2):597–603.

57. Allinquant B, Hantraye P, Mailleux P, Moya K, BouiUot C, Prochiantz A. Downregulation of amyloid precursor protein inhibits neurite outgrowth in vitro. *J Cell Biol* 1995;**128**(5):919–27.

58. Kamenetz F, Tomita T, Hsieh H, Seabrook G, Borchelt D, Iwatsubo T, et al. APP processing and synaptic function. *Neuron* 2003;**37**:925–37.

INTRACELLULAR CONSEQUENCES OF AMYLOID IN ALZHEIMER'S DISEASE

59. Abramov E, Dolev I, Fogel H, Ciccotosto GD, Ruff E, Slutsky I. Amyloid-β as a positive endogenous regulator of release probability at hippocampal synapses. *Nat Neurosci* 2009;**12**:1567–76.
60. Pearson HA, Peers C. Physiological roles for amyloid beta peptides. *J Physiol* 2006;**575**:5–10.
61. Mesulam MM. Review. Neuroplasticity failure in Alzheimer's disease: bridging the gap between plaques and tangles. *Neuron* 1999;**24**:521–9.
62. Hardingham GE, Bading H. Review. The Yin and Yang of NMDA receptor signalling. *Trends Neurosci* 2003;**26**:81–9.
63. Hardingham GE, Fukunaga Y, Bading H. Extrasynaptic NMDARs oppose synaptic NMDARs by triggering CREB shut-off and cell death pathways. *Nat Neurosci* 2002;**5**:405–14.
64. Papadia S, Stevenson P, Hardingham NR, Bading H, Hardingham GE. Nuclear Ca2 + and the cAMP response element-binding protein family mediate a late phase of activity-dependent neuroprotection. *J Neurosci* 2005;**25**:4279–87.
65. Mao Z, Bonni A, Xia F, Nadal-Vicens M, Greenberg ME. Neuronal activity-dependent cell survival mediated by transcription factor MEF2. *Science* 1999;**286**:785–90.
66. Alkon DL, Rasmussen H. A spatial-temporal model of cell activation. *Science* 1988;**239**:998–1005.
67. Zhang SJ, Zou M, Lu L, Lau D, Ditzel DA, Delucinge-Vivier C, et al. Nuclear calcium signaling controls expression of a large gene pool: identification of a gene program for acquired neuroprotection induced by synaptic activity. *PLoS Genet* 2009;**5**:e1000604.
68. Levin ED. Review. Nicotinic receptor subtypes and cognitive function. *J Neurobiol* 2002;**53**(4):633–40.
69. Wang H-Y, Lee DHS, D'Andrea MR, Peterson PA, Shank RP, Reitz AB. β-amyloid$_{1-42}$ binds to α7 nicotinic acetylcholine receptor with high affinity. *J Biol Chem* 2000;**275**(8):5626–32.
70. Dougherty JJ, Wu J, Nichols RA. Beta-amyloid regulation of presynaptic nicotinic receptors in rat hippocampus and neocortex. *J Neurosci* 2003;**23**:6740–7.
71. Cirrito JR, Yamada KA, Finn MB, Sloviter RS, Bales KR, May PC, et al. Synaptic activity regulates interstitial fluid amyloid-β levels in vivo. *Neuron* 2005;**48**:913–22.
72. Jia Y, Zhou J, Tai Y, Wang Y. TRPC channels promote cerebellar granule neuron survival. *Nat Neurosci* 2007;**10**:559–67.
73. Ramsden M, Plant LD, Webster NJ, Vaughan PF, Henderson Z, Pearson HA. Differential effects of unaggregated and aggregated amyloid beta protein (1–40) on K (+) channel currents in primary cultures of rat cerebellar granule and cortical neurones. *J Neurochem* 2001;**79**:699–712.
74. Ting JT, Kelley BG, Lambert TJ, Cook DG, Sullivan JM. Amyloid precursor protein overexpression depresses excitatory transmission through both presynaptic and postsynaptic mechanisms. *PNAS* 2007;**104**:353–8.
75. Priller C, Bauer T, Mitteregger G, Krebs B, Kretzschmar HA, Herms J. Synapse formation and function is modulated by the amyloid precursor protein. *J Neurosci* 2006;**26**:7212–21.
76. Cirrito JR, Kang JE, Lee J, Stewart FR, Verges DK, Silverio LM, et al. Endocytosis is required for synaptic activity-dependent release of amyloid-beta in vivo. *Neuron* 2008;**58**:42–51.
77. Reinhard C, Hebert SS, De Strooper B. The amyloid-beta precursor protein: integrating structure with biological function. *EMBO J* 2005;**24**:3996–4006.
78. Lesne S, Ali C, Gabriel C, et al. NMDA receptor activation inhibits α-secretase and promotes neuronal amyloid-β production. *J Neurosci* 2005;**25**(41):9367–77.
79. Brody DL, Magnoni S, Schwetye KE, Spinner ML, Esparza TJ, Stocchetti N, et al. Amyloid-β dynamics correlate with neurological status in the injured human brain. *Science* 2008;**321**:1221–4.

80. Kang JE, Lim MM, Bateman RJ, Lee JJ, Smyth LP, Cirrito JR, et al. Amyloid-β dynamics are regulated by orexin and the sleep-wake cycle. *Science* 2009;**326**:1005−7.
81. Rozmahel R, Huang J, Chen F, Liang Y, Nguyen V, Ikeda M, et al. Normal brain development in PS1 hypomorphic mice with markedly reduced γ-secretase cleavage of βAPP. *Neurobiol Aging* 2002;**23**:187−94.
82. Phinney AL, Drisaldi B, Schmidt SD, Lugowski S, Coronado V, Liang Y, et al. In vivo reduction of amyloid-β by a mutant copper transporter. *PNAS* 2003;**100**:14193−8.
83. Pawlik M, Sastre M, Calero M, Mathews PM, Schmidt SD, Nixon RA, et al. Overexpression of human cystatin C in transgenic mice does not affect levels of endogenous brain amyloid beta peptide. *J Mol Neurosci* 2004;**22**:13−18.
84. Mastrangelo P, Mathews PM, Chishti MA, Schmidt SD, Gu Y, Yang J, et al. Dissociated phenotypes in presenilin transgenic mice define functionally distinct γ-secretases. *PNAS* 2005;**102**:8972−7.
85. Puzzo D, Privitera L, Leznik E, Fa M, Staniszewski A, Palmeri A, et al. Picomolar amyloid-β positively modulates synaptic plasticity and memory in hippocampus. *J Neurosci* 2008;**28**:14537−45.
86. Cirrito JR, May PC, O'Dell MA, Taylor JW, Parsadanian M, Cramer JW, et al. In vivo assessment of brain interstitial fluid with microdialysis reveals plaque-associated changes in amyloid-β metabolism and half-life. *J Neurosci* 2003;**23**:8844−53.
87. Trinchese F, Liu S, Ninan I, Puzzo D, Jacob JP, Arancio O. Cell cultures from animal models of Alzheimer's disease as a tool for faster screening and testing of drug efficacy. *J Mol Neurosci* 2004;**24**:15−21.
88. Garcia-Osta A, Alberini CM. Amyloid beta mediates memory formation. *Learn Mem* 2009;**16**:267−72.
89. Wu J, Anwyl R, Rowan MJ. Beta-amyloid selectively augments NMDA receptor-mediated synaptic transmission in rat hippocampus. *Neuroreport* 1995;**6**:2409−13.
90. Wu J, Anwyl R, Rowan MJ. Beta-amyloid-(1−40) increases long-term potentiation in rat hippocampus in vitro. *Eur J Pharm* 1995;**284**:R1−3.
91. Koudinov AR, Berezov TT. Review. Alzheimer's amyloid-β (Aβ) is an essential synaptic protein, not neurotoxic junk. *Acta Neurobiol Exp* 2004;**64**:71−9.
92. Schulz PE. Beta-peptides enhance the magnitude and probability of long term potentiation. *Soc Neurosci Abstr* 1996;**22**:2111.
93. Muller U, Cristina N, Li ZW, Wolfer DP, Lipp HP, Rulicke T, et al. Behavioral and anatomical deficits in mice homozygous for a modified β-amyloid precursor protein gene. *Cell* 1994;**79**:755−65.
94. Dawson GR, Seabrook GR, Zheng H, Smith DW, Graham S, O'Dowd G, et al. Age-related cognitive deficits, impaired long-term potentiation and reduction in synaptic marker density in mice lacking the β-amyloid precursor protein. *Neuroscience* 1999;**90**:1−13.
95. Seabrook GR, Smith DW, Bowery BJ, Easter A, Reynolds T, Fitzjohn SM, et al. Mechanisms contributing to the deficits in hippocampal synaptic plasticity in mice lacking amyloid precursor protein. *Neuropharmacology* 1999;**38**:349−59.
96. Luo L, Tully T, White K. Human amyloid precursor protein ameliorates behavioral deficit of flies deleted for Appl gene. *Neuron* 1992;**9**:595−605.
97. Mileusnic R, Lancashire CL, Johnston AN, Rose SP. APP is required during an early phase of memory formation. *Eur J Neurosci* 2000;**12**:4487−95.
98. Doyle E, Bruce MT, Breen KC, Smith DC, Anderton B, Regan CM. Intraventricular infusions of antibodies to amyloid-β−protein precursor impair the acquisition of a passive avoidance response in the rat. *Neurosci Lett* 1990;**115**:97−102.
99. Huber G, Martin JR, Loffler J, Moreau JL. Involvement of amyloid precursor protein in memory formation in the rat: an indirect antibody approach. *Brain Res* 1993;**603**:348−52.

100. Squire LR. Review. Memory and the hippocampus: a synthesis from findings with rats, monkeys, and humans. *Psychol Rev* 1992;**99**:195—231.

101. Meziane H, Dodart JC, Mathis C, Little S, Clemens J, Paul SM, et al. Memory-enhancing effects of secreted forms of the β-amyloid precursor protein in normal and amnestic mice. *PNAS* 1998;**95**:12683—8.

102. Chin JH, Ma L, MacTavish D, Jhamandas JH. Amyloid β protein modulates glutamate-mediated neurotransmission in the rat basal forebrain: involvement of pre-synaptic neuronal nicotinic acetylcholine and metabotropic glutamate receptors. *J Neurosci* 2007;**27**:9262—9.

103. D'Andrea MR, Lee DHS, Wang H-Y, Nagele RG. Targeting intracellular Aβ42 for Alzheimer's disease drug discovery. *Drug Dev Res* 2002;**56**:194—200.

104. D'Andrea MR, Nagele RG. Targeting the alpha 7 nicotinic acetylcholine receptor to reduce amyloid accumulation in Alzheimer's disease pyramidal neurons. *Curr Pharm Des* 2006;**12**:677—84.

105. Nagele RG, D'Andrea MR, Anderson WJ, Wang H-Y. Intracellular accumulation of β-amyloid$_{1-42}$ in neurons is facilitated by the α7 nicotinic acetylcholine receptor in Alzheimer's disease. *Neuroscience* 2002;**110**(2):199—211.

106. Morley JE, Farr SA, Banks WA, Johnson SN, Yamada KA, Xu L. A physiological role for amyloid-β protein: enhancement of learning and memory. *J Alzheimers Dis* 2010;**19**(2):441—9.

107. Wang H, Megill A, He K, Kirkwood A, Lee H-K. Review. Consequences of inhibiting amyloid precursor protein processing enzymes on synaptic function and plasticity. *Neural Plast* 2012;**272374**:24 pp.

108. Laird FM, Cai H, Savonenko AV, Farah MH, He K, Melnikova T, et al. BACE1, a major determinant of selective vulnerability of the brain to amyloid-β amyloidogen-esis, is essential for cognitive, emotional, and synaptic functions. *J Neurosci* 2005;**25**(50):11693—709.

109. Vassar R. Review. BACE1: the beta-secretase enzyme in Alzheimer's disease. *J Mol Neurosci* 2004;**23**(1—2):105—14.

110. Ohno M, Sametsky EA, Younkin LH, Oakley H, Younkin SG, Citron M, et al. BACE1 deficiency rescues memory deficits and cholinergic dysfunction in a mouse model of Alzheimer's disease. *Neuron* 2004;**41**:27—33.

111. Hoffmann J, Twiesselmann C, Kummer MP, Romagnoli P, Herzog V. A possible role for the Alzheimer amyloid precursor protein in the regulation of epidermal basal cell proliferation. *Eur J Cell Biol* 2000;**79**:905—14.

112. Graebert KS, Lemansky P, Kehle T, Herzog V. Localization and regulated release of Alzheimer amyloid precursor-like protein in thyrocytes. *Lab Invest* 1995;**72**:513—23.

113. Almeida CG, Takahashi RH, Gouras GK. Beta-amyloid accumulation impairs multi-vesicular body sorting by inhibiting the ubiquitin-proteasome system. *J Neurosci* 2006;**26**(16):4277—88.

114. Naves FJ, Calzada B, Cabal A, Alonso-Cortina V, Del Valle ME, Fermindez-Sanehez MT, et al. Expression of flamyloid precursor protein (APP) in human dorsal root ganglia. *Neurosci Lett* 1994;**181**:73—7.

115. Schmitt TL, Steiner E, Klingler P, Lassmann H, Grubeck-Loebenstein B. Thyroid epi-thelial cells produce large amounts of the Alzheimer β-amyloid precursor protein (APP) and generate potentially amyloidogenic APP fragments. *J Clin Endocrinol Metab* 1995;**80**(12):35139.

116. Galloway S, Jian L, Johnsend R, Chewa S, Mamob JCL. Beta-amyloid or its precursor protein is found in epithelial cells of the small intestine and is stimulated by high-fat feeding. *J Nutr Biochem* 2007;**18**:279—84.

117. Galloway S, Pallebage-Gamarallage M, Takechi R, Jian L, Johnsen RD, Dhaliwal SS, et al. Synergistic effects of high fat feeding and apolipoprotein E deletion on entero-cytic amyloid-β abundance. *Lipids Health Dis* 2008;**7**:15.

118. Schmitz A, Tikkanen R, Kirfel G, Herzog V. The biological role of the Alzheimer amyloid precursor protein in epithelial cells. *Histochem Cell Biol* 2002;**117**:171–80.

119. Boyt AA, Taddei K, Hallmayer J, Mamo J, Helmerhorst E, Gandy SE, et al. Relationship between lipid metabolism and amyloid precursor protein and apolipoprotein E. *Alzheimers Rep* 1999;**2**:339–46.

120. Kalmijn S, Launer LJ, Ott A, Witteman JC, Hofman A, Breteler MM. Dietary fat intake and the risk of incident dementia in the Rotterdam Study. *Ann Neurol* 1997;**42**:776–82.

121. Solfrizzi V, D'Introno A, Colacicco AM, Capurso C, Del Parigi A, Capurso S, et al. Dietary fatty acids intake: possible role in cognitive decline and dementia. *Exp Gerontol* 2005;**40**:257–70.

122. Sparks DL, Scheff SW, Hunsaker III JC, Liu H, Landers T, Gross DR. Induction of Alzheimer-like β-amyloid immunoreactivity in the brains of rabbits with dietary cholesterol. *Exp Neurol* 1994;**126**:88–94.

123. Refolo LM, Malester B, LaFrancois J, Bryant-Thomas T, Wang R, Tint GS, et al. Hypercholesterolemia accelerates the Alzheimer's amyloid pathology in a transgenic mouse model. *Neurobiol Dis* 2000;**7**:321–31.

124. Shie FS, Jin LW, Cook DG, Leverenz JB, LeBoeuf RC. Diet-induced hypercholesterolemia enhances brain Aβ accumulation in transgenic mice. *Neuroreport* 2002;**13**:455–9.

125. Caille I, Allinquant B, Dupont E, Bouillot C, Langer A, Muller U, et al. Soluble form of amyloid precursor protein regulates proliferation of progenitors in the adult subventricular zone. *Development* 2004;**131**:2173–81.

126. Popp GM, Graebert KS, Pietrzik CU, Rosentreter SM, Lemansky P, Herzog V. Growth regulation of rat thyrocytes (FRTL-5 cells) by the secreted ectodomain of β-amyloid precursor-like proteins. *Endocrinology* 1996;**137**:1975–83.

127. Ishiguro M, Ohsawa I, Takamura C, Morimoto T, Kohsaka S. Secreted form of β-amyloid precursor protein activates protein kinase C and phospholipase Cγ1 in cultured embryonic rat neocortical cells. *Brain Res Mol Brain Res* 1998;**53**:24–32.

128. Cheng G, Yu Z, Zhou D, Mattson MP. Phosphatidylinositol-3-kinase-Akt kinase and p42/p44 mitogen-activated protein kinases mediate neurotrophic and excitoprotective actions of a secreted form of amyloid precursor protein. *Exp Neurol* 2002;**175**:407–14.

129. Mook-Jung I, Saitoh T. Amyloid precursor protein activates phosphotyrosine signaling pathway. *Neurosci Lett* 1997;**235**:1–4.

130. Bolos M, Spuch C, Ordoñez-Gutierrez L, Wandosell F, Ferrer I, Carro E. Neurogenic effects of β-amyloid in the choroid plexus epithelial cells in Alzheimer's disease. *Cell Mol Life Sci* 2013;**70**:2787–97.

131. López-Toledano MA, Shelanski ML. Neurogenic effect of β-amyloid peptide in the development of neural stem cells. *J Neurosci* 2004;**24**:5439–44.

132. Calafiore M, Copani A, Deng W. DNA polymerase-β mediates the neurogenic effect of β-amyloid protein in cultured subventricular zone neurospheres. *J Neurosci Res* 2010;**90**:559–67.

133. Mattson MP. Review. Cellular actions of β-amyloid precursor protein and its soluble and fibrillogenic derivatives. *Physiol Rev* 1997;**77**:1081–132.

134. Cao X, Sudhof TC. A transcriptionally active complex of APP with Fe65 and histone acetyltransferase Tip60. *Science* 2001;**293**:115–20.

135. Scheinfeld MH, Ghersi E, Laky K, Fowlkes BJ, D'Adamio L. Processing of β-amyloid precursor-like protein-1 and -2 by γ-secretase regulates transcription. *J Biol Chem* 2002;**277**(46):44195–201.

136. Soscia SJ, Kirby JE, Washicosky KJ, Tucker SM, Ingelsson M, Hyman B, et al. The AD-associated amyloid β-protein is an antimicrobial peptide. *PLoS One* 2010;**5**(3): e9505.

137. Imbimbo BP. Review. Why did tarenflurbil fail in AD? *J Alzheimers Dis* 2009;**17**(4):757–60.
138. Green RC, Schneider LS, Amato DA, Beelen AP, Wilcock G, et al. Effect of tarenflurbil on cognitive decline and activities of daily living in patients with mild Alzheimer disease: a randomized controlled trial. *JAMA* 2009;**302**:2557–64.
139. Putsep K, Carlsson G, Boman HG, Andersson M. Deficiency of antibacterial peptides in patients with morbus Kostmann: an observation study. *Lancet* 2002;**360**:1144–9.
140. Dominguez D, Tournoy J, Hartmann D, Huth T, Cryns K, et al. Phenotypic and biochemical analyses of BACE1- and BACE2-deficient mice. *J Biol Chem* 2005;**280**:30797–806.
141. Collin RWJ, van den Hurk WH, Martens GJM. Biosynthesis and differential processing of two pools of amyloid-β precursor protein in a physiologically inducible neuroendocrine cell. *J Neurochem* 2005;**94**:1015–24.
142. Jenks BG, Leenders HJ, Martens GJ, Roubos EW. Adaptation physiology: the functioning of pituitary melanotrope cells during background adaptation of the amphibian *Xenopus laevis*. *Zool Sci* 1993;**10**:1–11.
143. Martens GJ, Weterings KA, van Zoest ID, Jenks BG. Physiologically-induced changes in proopiomelanocortin mRNA levels in the pituitary gland of the amphibian *Xenopus* laevis. *Biochem Biophys Res Commun* 1987;**143**:678–84.
144. Kamal A, Stokin GB, Yang Z, Xia C-H, Goldstein LBS. Axonal transport of amyloid precursor protein is mediated by direct binding to the kinesin light chain subunit of kinesin-I. *Neuron* 2000;**28**:449–59.
145. Muller U, Kins S. APP on the move. *Trends Mol Med* 2002;**8**:152–5.
146. Kummer C, Wehner S, Quast T, Werner S, Herzog V. Expression and potential function of β-amyloid precursor proteins during cutaneous wound repair. *Exp Cell Res* 2002;**280**(2):222–32.
147. Li QX, Whyte S, Tanner JE, Evin G, Beyreuther K, Masters CL. Secretion of Alzheimer's disease Aβ amyloid peptide by activated human platelets. *Lab Invest* 1998;**78**:461–9.
148. Morris GP, Clark IA, Vissel B. Review. Inconsistencies and controversies surrounding the amyloid hypothesis of AD. *Acta Neuropathol Commun* 2014;**2**:135.
149. Senechal Y, Larmet Y, Dev KK. Unraveling in vivo functions of amyloid precursor protein: insights from knockout and knockdown studies. *Neurodegener Dis* 2006;**3**:134–47.
150. Turner PR, O'Connor K, Tate WP, Abraham WC. Review. Roles of amyloid precursor protein and its fragments in regulating neural activity, plasticity and memory. *Prog Neurobiol* 2003;**70**(1):1–32.
151. Ramsden M, Nyborg AC, Murphy MP, Chang L, Stanczyk FZ, Golde TE, et al. Androgens modulate β-amyloid levels in male rat brain. *J Neurochem* 2003;**87**:1052–5.

Pathological Consequences of Aβ From Extracellular to Intraneuronal

Aβ is not only associated with AD, but also one of the primary causes of the disease as supported by the hundreds of publications and monetary support for the possibility. The evidence presented through the 1980s and 1990s provided enough justification to test the amyloid hypothesis in humans.

Intracellular Consequences of Amyloid in Alzheimer's Disease.
DOI: http://dx.doi.org/10.1016/B978-0-12-804256-4.00004-8

69

EXTRACELLULAR Aβ: THE AMYLOID CASCADE HYPOTHESIS

The presence of Aβ in the senile plaques in ADs brain tissues is one of the cornerstones of the "amyloid cascade hypothesis," and has driven drug development strategies for over 20 years.[1,2] This hypothesis emphasizes that increased neuronal Aβ production and secretion leading to extracellular accumulation of Aβ in the brain over time results in the formation of amyloid plaques, which induce inflammatory responses, and in turn induce synaptic damage, neurofibrillary tangles (NFTs), neuronal loss, and then AD (Fig. 4.1).[1-4]

There are several building blocks that lead to the development of the amyloid cascade hypothesis: genetic support (eg, Down syndrome, gene mutations in APP-related molecules, *APOE4* genotype, and transgenic preclinical models), extracellular aggregation of toxic Aβ resulting in senile plaques and inflammation, and intracellular NFTs.

Genetic Support

APP Processing Mutations

Some of the early evidence that led to the formulation of the amyloid hypothesis originated from the AD sufferers with the FAD mutations, who have increased production of Aβ with increased accumulations of Aβ in their brain that also present an early onset of AD.[5-9] Genetic linkage studies indicate that a small portion of individuals with AD have an autosomal dominant pattern of inheritance with three mutations that affect the metabolism or stability of Aβ: APP, presenilin 1 (PSEN1, or PS1), and PSEN2 (or PS2).[5,6,10]

APP is a type-1 transmembrane glycoprotein with 10 isoforms generated by alternative mRNA splicing, which is encoded by a single gene on the human chromosome 21.[11-13] APP undergoes several pathways of processing to yield products including the Aβ species (see Chapter 1). Missense mutations (defined as a point mutation whereby a single nucleotide change results in a codon that codes for a different amino acid) have been reported on the APP gene in families who have FAD.[14,15]

Of the many reported APP gene mutations (Fig. 4.2), one common mutation in APP is known as the Swedish FAD mutation in which a double amino acid change (Lys670Asn/Met671Leu) leads to increased cleavage of APP by the β-secretase and therefore, more production of Aβ.[16] Increased Aβ production, especially the generation of Aβ42, also occurs with the Flemish mutation (Ala693Gly), as well as in the Dutch mutation (Glu693Gln).[6,9,17] The changes in Aβ peptide sequence resulting

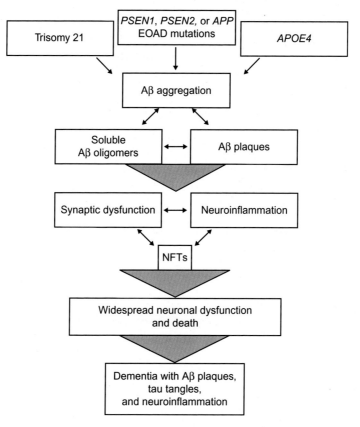

FIGURE 4.1 The amyloid hypothesis. The amyloid hypothesis postulates that Aβ aggregation triggers a cascade of events ultimately resulting in AD. Familial mutations in *PSEN1*, *PSEN2*, or *APP* are associated with EOAD. These genetic risk factors are postulated to impact the cleavage of Aβ from APP, leading to oligomerization and eventual Aβ plaque formation. Individuals with trisomy 21 (Down syndrome), and therefore a triple copy of APP, suffer EOAD. The strongest genetic risk factor for LOAD is the presence of at least one *APOE4* allele. It is unclear as to what triggers Aβ accumulation in LOAD, though it is suggested that there may be a number of contributing factors such as reduced Aβ clearance due to *APOE* genotype. Aβ oligomerization is proposed to trigger a cascade involving the formation of NFTs composed of hyperphosphorylated tau, synapse loss, neuron death, and widespread neuroinflammation, particularly in brain regions involved in learning and memory, such as the hippocampus. As the amyloid burden increases, the ongoing catastrophic loss of synapses and neurons is thought to lead to progressive dementia. *Aβ*, amyloid-β; *APP*, amyloid precursor protein; *APOE4*, apolipoprotein E4; *EOAD*, early-onset Alzheimer's disease; *LOAD*, late-onset Alzheimer's disease; *NFT*, neurofibrillary tangles; *PSEN*, presenilin. *Source: Used with permission from* Acta Neuropathol Commun *2014;2:135.*

(A)

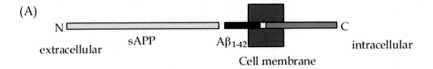

(B)

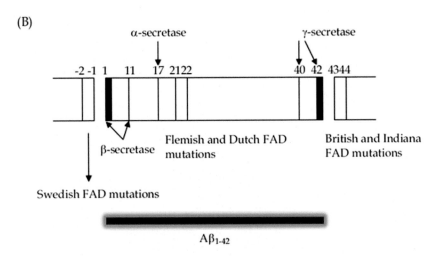

FIGURE 4.2   Schematic diagrams of APP and its cleavage. (A) Full-length APP is located in the cell membrane. (B) APP can be cleaved at α-, or β-, and γ-secretase sites. FAD mutations are often at cleavage sites. *APP*, amyloid precursor protein. *Source: Used with permission from* Advanced Understanding of Neurodegenerative Diseases, *2011.*

from the Dutch mutation leads to a conformational change that leads to increases in Aβ aggregation and forms fibrils.[17,18] The Flemish mutation also affects the Aβ peptide sequence by altering the γ-secretase activity leading to increased production of Aβ42.[19] The British and Indiana mutations also increase Aβ peptide aggregation.[6] Other mutations, such as the Arctic mutation (APPArc) do not increase Aβ production, but do increase the aggregation of Aβ leading to an early onset, aggressive forms of the disease.[20]

Although mutations in the APP gene can also lead to the development of early onset AD by altering/increasing the proteolytic cleavage of APP, most of the FAD cases are linked to point mutations in the presenilin genes.[5,10,21] Presenilins are transmembrane proteins with eight transmembrane domains located mainly in the ER, Golgi, ER-Golgi intermediate structures, and synaptic terminals as detected by electron microscopy.[3] Mutations in the presenilins genes, presenilin 1 on human chromosome 14 and presenilin 2 on chromosome 1, in subjects with

FAD also increase Aβ production and aggregation. If an individual carries a presenilin mutation, the probability of developing early onset AD is higher than 95%. The majority of these presenilin mutations are missense mutations leading to amino acid change in the protein sequence. For example, the PS1M146V mutation increases levels of Aβ42, which aggregates more readily than Aβ40.[22,23]

Down Syndrome

Additional support for the critical role of Aβ in the amyloid cascade hypothesis comes from individuals with Down syndrome. The earliest connection between patients with AD and Down syndrome came when the amyloid protein blamed for AD was homologous to that in Down syndrome.[24,25] It was determined that the APP gene located on chromosome 21, which is in triplicate in Down syndrome patients, was responsible for the increases in the production of Aβ early in life.[15,16,24,26−29] Compared with age-matched controls from the general population, the Aβ42 and Aβ40 plasma concentrations were higher in adults with Down syndrome.[30−32] Mean plasma levels of Aβ42 were higher in Down syndrome subjects with dementia as compared with nondemented Down syndrome subjects of similar ages.[32] Whether the increased concentrations of Aβ are the cause or effect, all Down syndrome patients will inevitably develop AD pathology (eg, diffuse and neuritic/senile plaques) in their fourth decade of life.[33−36] The majority of Down syndrome patients will develop dementia by their seventies, thereby providing compelling evidence that the increased expression of APP could facilitate AD progression.[35]

Transgenic AD Preclinical Models

Given the lack of definitive biomarkers of AD in humans, transgenic animal models of the amyloid-based pathology based on FAD continue to be valuable tools to examine molecular changes preceding the deposition of amyloid plaques and associated pathology (ie, late inflammation, neuritic dystrophy, etc.).[37−39] As with many animal models of human disease, it is critical to validate the model to determine how tightly linked the model mimics the human condition, and in the instance of AD, it would be cognitive decline.

Behavioral tests are used to track progressive cognitive decline, which is primarily due to the loss of neurons and synapses in the hippocampal formation and related areas.[40] Such changes in cognitive decline in the mouse or rat models of AD are typically divided into either associative or operant learning tasks. The former uses cues in the environment to condition a specific spatial response in the animal, while the later requires the animal to make a particular response to a stimulus in order to receive an outcome. Of the types of memory tasks, the spatial memory task includes the Morris water maze, radial arm maze, the radial arm water maze, and the contextual memory tasks include fear conditioning, passive-avoidance learning. Another type of behavioral task tests animals for their working memory based on novelty and activity in the Y-maze, T-maze, object recognition, and open field tasks.

Mouse

The first transgenic mouse model of AD utilized the platelet-derived growth factor promoter to overexpress APP. These mice display increases in human Aβ40 and Aβ42 by as much as 5—14 times the concentration in the wild-type mouse.[40]

The next transgenic mouse model included the human mutant APP transgenic mouse model, named Tg2576, that overexpressed the Swedish double mutant form of APP695. This model produces five times the endogenous levels of APP in the brain, and after 11 months, it develops other neuropathological features including plaques of extracellular Aβ42 and Aβ40 deposits.

The role of the presenilins was investigated in the PS1 knockout mouse. Unfortunately, they quickly died after birth. However, of the few types of PS1 and PS2 transgenic mouse models that survived, all demonstrated cognitive decline (eg, altered synaptic transmission and hippocampal LTP) without accompanying plaque-like accumulations or behavioral alterations.[41,42]

A multiple gene transgenic mouse model of FAD was developed that altered both the presenilins and the accumulation of human APP named APP + PS1 mice. This was one of the first transgenic mouse models to show a strong positive correlation between Aβ42 development and cognitive decline.[40]

Impairment of basal synaptic transmission occurs in a variety of other APP transgenic mouse models.[43–45] In addition, deficits in induction of long-term synaptic changes are also observed in several APP transgenic lines, like PDAPP mice, APP London, Tg2576, and in the triple-transgenic mouse model.[46–49]

Rat

Although the mouse models of FAD provide investigators opportunities to assess the effects of excessive Aβ on behavior and pathology, a rat model was introduced as a more desirable transgenic model of AD because of the extensive Neuroscience Database and the richer behavioral display as compared to mice.[50] The larger brain size also makes surgical procedures and pharmacological manipulation easier to perform. Some transgenic rat models were developed but failed to produce extracellular deposition of amyloid plaques. One of these models presented intraneuronal accumulation of Aβ coupled with changes in subcellular organelles, alterations in the hippocampal proteomic patterns, and cognitive impairments.

A triple-transgenic rat was developed, which co-expressed three transgenes: human AβPP Swedish mutation, human AβPP Swedish and Indiana mutations, and human presenilin 1 mutation. These triple-transgenic rats developed extracellular deposition of mature plaques by 9 months of age, and showed cognitive impairment as early as 2 months, which was prior to the presence of extracellular amyloid plaques.

Another transgenic rat model of AD was the generation of the McGill-RThy1-APP rat. This transgenic model expresses a single trans-gene coding for a modified variant of the human protein AβPP751, containing both the Swedish and Indiana mutations. A full AD-like amyloid pathology, including the generation of mature plaques and dystrophic neurites, is presented in this model.[37,50]

Closing Comment

None of the transgenic models described so far exhibits all of the AD features.[38] Although investigating these genetic conditions are critical in deducing the biological pathogenesis of AD, it must be remembered that animal models genetically engineered to have mutations in these three genes are important in assessing their role in AD but represent the minority. These APP-related mutations are responsible for only 30–50% of FAD cases, which in turn only represents 6–8% of all AD cases suggesting that many other factors are involved in disease pathology.[5,10,51] Therefore, the availability of human tissue to study sporadic cases of AD is invaluable in helping to understand the pathological processes leading to AD.

APOE ε4 Allele

The most prominent genetic risk factor for AD is the gene that codes for *APOE ε4 (APOE4)*.[52–59] These lipoproteins are responsible for packaging cholesterol and other fats, and for transporting them through the bloodstream. APOE is also a major component of a specific type of lipoprotein, so-called very-low-density lipoproteins, which remove excess cholesterol from the blood to the liver for processing. Maintaining normal levels of cholesterol is essential for the prevention of disorders of the cardiovascular system, including hypertension, heart attack, stroke, and hypercholesterolemia, all of which are AD risk factors.[60]

The *APOE2* and *APOE3* gene forms are most common in the general population. The *APOE4* gene is associated with an individual's risk for developing late-onset AD as evident by the greater accumulation of Aβ in the elderly with and without AD with a frequency of ~15% in general populations but >50% in AD patients.[53,61] Inheritance of one or two copies of the *ε4* allele is associated with a dose-dependent increased risk for AD and an earlier age of onset of AD characterized with a higher depositions of Aβ42 and Aβ40 in the brains of sporadic AD cases (ie, load of senile plaques) and cerebral amyloid angiopathy (CAA).[53,62–70] This was most evident in patients between the ages of 50 and 59 years with ~41% of *ε4*-carriers bearing senile plaques, compared to only ~8% in noncarriers.[71] *APOE4* influences β-amyloid degradation, brain, and neuronal activity.[62,72] In vitro studies show that APOE is able to bind to Aβ, and APOE may also accelerate the formation of Aβ fibrils.[65,73–75] There is a differential development of AD-associated changes in the brain of individuals having at least one *ε4* allele.

Similarly, patients with Down syndrome who have the *APOE ε4* allele were associated with earlier onset of dementia, as compared to the general population, while the *APOE ε2* allele was highly protective.[76,77] The distribution of *APOE* alleles was consistent with that found previously in Down syndrome and in the general population (*ε2*, 7.3%; *ε3*, 78.9%; and *ε4*, 13.7%).[64,77] Higher mean plasma Aβ42 concentrations were observed in Down syndrome patients with at least one *ε4* allele, independent of mental status as compared to patients without the *ε4* allele. Higher mean plasma Aβ42 concentrations were higher in Down syndrome patients with dementia as compared with those nondemented patients.[32] The *ε4* allele was associated with increased plasma levels of Aβ42 but not Aβ40. The selective effect of the *ε4* allele on elevation of plasma Aβ42 levels was consistent with a mechanism of action involving acceleration of the rate of amyloid fibril formation or diminished clearance of Aβ.[66,68] These data support the hypothesis that Aβ42 plays an important role in pathogenic progression in dementia in Down syndrome.[32]

Toxic Extracellular Aβ

Extracellular Aβ deposition, a major pathological hallmark of AD, is an essential component of the amyloid cascade hypothesis, which positions the toxic extracellular Aβ, especially extracellular Aβ42, as one of the primary causes of neuronal death and AD. While the Aβ40 is the most common form, it is the Aβ42 that is the most fibrillogenic, and believed to be associated with FAD and memory disorders. The extracellular Aβ toxicity-based amyloid cascade hypothesis is supported by the evidence that fibrillar extracellular Aβ is toxic to various systems including cell lines and primary cells in cultures. Levels of Aβ, especially Aβ42, increase in the AD brains, the serum, and fibroblasts from the AD patients.

These deposits of toxic Aβ are thought to seed or accumulate into the senile (also known as neuritic or dense-core) amyloid plaques. Although the extracellular amyloid plaques are the signature of amyloid cascade hypothesis, a myriad of molecular changes at the cellular level have also been attributed to extracellular Aβ (Table 4.1), which may be triggered differentially during the progress of the disease that most probably contribute to different stages of the pathology.[3,36]

Intracerebral administration of high concentrations (nM−μM range) of Aβ peptides, which mimic Aβ accumulation in AD, disrupted the retention of both spatial and contextual fear memories, as well as short-term working memory in rodents.[78−81] The soluble Aβ oligomers rather than large Aβ aggregates or deposits were responsible for disrupting the mechanisms underlying learning and memory, particularly during the earliest stages of AD. In fact, the accumulation of Aβ oligomers in the brain of normal rats also altered memory LTP and LTD and impaired memory retention.[79,82−84]

Based on these and many other reports, extracellular Aβ, especially Aβ42 is neurotoxic, and can lead to the demise of neurons to eventually produce the clinical symptoms presented in AD patients. However, this presumption may be inaccurate because in spite of lowering levels of extracellular Aβ as well as the number of amyloid plaques in patients, cognitive impairment in the AD patients remained unchanged.

TABLE 4.1 Examples of Reported Toxic Effects of Extracellular Aβ[3,36]

- Activation of inflammation and microglia
- Alterations in the neuronal endosomal-lysosomal system; induction of lysosomal protease activity and damaging membrane
- Calcium homeostasis; changes in calcium influx
- Changes in tau phosphorylation
- Gene expression
- Increasing vulnerability of cells to a secondary insult
- Induction of apoptosis
- Mitochondrial respiration
- Neurotransmitter modulation
- Proteasome activity
- Protein kinases activity
- The complement cascade
- The redox status of the cell/increasing oxidative stress

INTRACELLULAR CONSEQUENCES OF AMYLOID IN ALZHEIMER'S DISEASE

Revised Hypothesis

The failure to improve cognitive function in patients with AD prompted some investigators to revise the amyloid cascade hypothesis. The initial postulate that Aβ deposition represents the critical event that prompts neurodegeneration had to be reconsidered when it was found that Aβ brain load did not correlate with neuronal loss and cognitive impairment, whereas tau pathology could more closely reflect the disease progression with NFTs correlating to neuronal loss and cognitive deficits.[4,85]

Subsequently, the amyloid hypothesis was revised to state that it is the soluble Aβ oligomers that represent the toxic species that trigger neuronal death or/and synaptic dysfunction, and possibly, that a threshold level of Aβ has to be reached to induce toxicity.[86] This means that Aβ production is an early event in the disease process that precedes the appearance of plaques and tangles, and also precedes clinical symptoms and cognitive deficits caused by neuronal loss. Identifying early changes in APP and secretase activities in platelets may indicate similar changes occurring in the brain that will lead to a patient's progression from preclinical stage to MCI and AD.[87]

Architects of the amyloid hypothesis remain steadfast that the litany of studies has supported the hypothesis in its "broad outlines."[88] The development of anti-Aβ therapeutics remains the most rational approach in treating AD,[88] while some suggest the reason for the hypothesis failure is that the disease is not being targeted early enough.

INTRACELLULAR Aβ: Aβ'S LAST STAND

Inaccuracies in the Once Widely Accepted Amyloid Hypothesis

The survival of the amyloid cascade hypothesis in either configuration has been extraordinary, and although the amyloid hypothesis has shifted its focus from plaque fibrillar-amyloid to soluble forms of Aβ over the recent years, it largely remains defined by the central tenet that extracellular accumulation of amyloid in a variety of forms triggers a cascade that harms neurons and synapses.[2]

To quote a source: "Rather than considering that the unreliability of plaque as a disease marker that may reflect badly on the amyloid hypothesis, its guardians embraced soluble Aβ oligomers as a cause of AD."[89] Many have concerns over revisions to the amyloid hypothesis. One suggested that the amyloid hypothesis now concerns "...invisible molecules that target invisible structures."[90] New information is only interpreted within a constantly fluctuating amyloid hypothesis rather than being reconfigured into alternative hypothesis, which may better explain disease causality.

The failure of the clinical trials triggered some scientists to question the legitimacy of the amyloid hypothesis, and whether lowering amyloid concentrations via vaccination, secretase modulation, or may any other means could ever be beneficial.[91] Those failures left some investigators to wonder if the hypothesis will ever die, as it has been so modified over time that it is now impossible to confirm or deny.[90] The hypothesis has become "so difficult to challenge because so often it is the lens through which peer reviewers, granting bodies, and pharmaceuticals judge or support AD research."[2] Nonamyloid theory data presented to challenge the hypothesis tends to be overlooked and integrated within the amyloid hypothesis, leaving such work suffocated.

In spite of the dismal results from each anti-Aβ clinical trial, new anti-Aβ trials are initiated "with no hint of subsiding."[90] The putative Alzheimer's curing-hypothesis has gradually become a dogma, even as far back as in the early 2000's, "The current dogma posits that Aβ42 accumulates extracellularly, leading to the formation of amyloid plaques."[91–94]

Many report inconsistencies in the amyloid hypothesis (Fig. 4.3).[2,3,95] For example, the notion that extracellular Aβ leads to AD is not accurate. In AD patients, the severity of Aβ deposition correlates poorly with clinical dementia levels. In some AD animal models, Aβ accumulates and forms senile plaques in the absence of the other two AD pathological features, neuronal loss and NFTs. Extracellular Aβ toxicity generally requires nonphysiological μM levels of Aβ in the culture medium. While others state that extracellular Aβ is not toxic even at high μM concentration in rat PC12, human IMR32 cells, and in monkey cerebral cortex. Furthermore, older people without dementia have numerous senile plaques in their brains.

Stereologic and image analyses revealed substantial age-related neuron loss in the hippocampal pyramidal cell layer of APP/PS1 double-transgenic mice.[96] The loss of neurons was observed at sites of Aβ aggregation and surrounding astrocytes but, most importantly, was also clearly observed in areas of the parenchyma distant from plaques. These findings suggest the insignificant involvement of extracellular Aβ in the loss of hippocampal neuron loss in this APP/PS1 double-transgenic mouse model of AD. Additionally, extracellular amyloid was rarely observed around normal or degenerating neurons.[97]

New Hypotheses

Issues with the well-funded dogmatic orthodoxy fueled the need for new approaches. It took the failure of the clinical trials to soundly refute a hypothesis, which some referred to as "at very least, premature."[2]

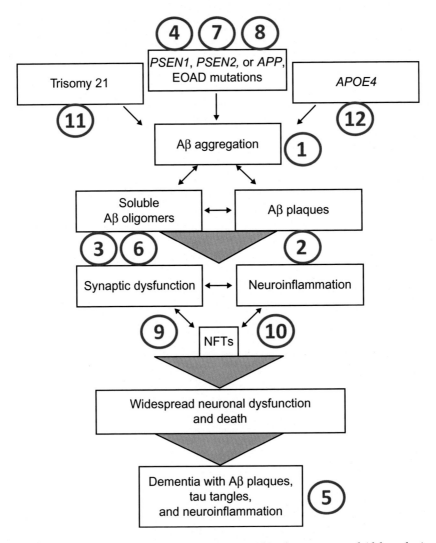

FIGURE 4.3 Controversies and inconsistencies within the current amyloid hypothesis. (1) Aβ deposition occurs in cognitively normal individuals; (2) There is a weak correlation between plaque load and cognition; (3) The biochemical nature and presence of Aβ oligomeric assemblies in vivo is unclear; (4) Preclinical AD models based on FAD-linked mutations are biased toward the amyloid hypothesis; (5) Pathological heterogeneity and comorbidities are unexplained by the amyloid hypothesis; (6) Aβ has a normal physiological role and targeting Aβ may disrupt these roles over the long term; (7) Genetic factors linked to AD can be interpreted independently of amyloid; (8) APP cleavage and function is more complex than solely the production of Aβ, indicating other APP family members may play a role in disease progression; (9) The triggers of synapse loss, neuronal loss, and neuroinflammation in AD are still unclear; (10) The relationship between Aβ and tau pathologies is unclear; (11) The onset of dementia in Down syndrome is highly variable,

One approach positioned the detection of intraneuronal amyloid into a new hypothesis that was presented in 1999, well before the bleak results of the Aβ-based clinical trials were known.[98] This hypothesis was based on the observations showing significantly higher levels of intraneuronal Aβ42 in the brain of AD brains as compared with age-matched controls, which was referred to as the "Inside-Out hypothesis" in 2001.[92,97,99,100] The origin of the hypothesis name was to convey how dense-core, senile, and neuritic plaques form from excessive accumulations of Aβ "inside" the neurons that eventually burst or lyse releasing its neuronal cargo "out" into the extracellular spaces in the brain. This simple concept focuses on intraneuronal rather than extracellular Aβ.

Recently, more and more scientists report pathological links between levels of intraneuronal Aβ and AD, which is the purpose of the remainder of this book. For example, some propose an extension of modification to the original amyloid hypothesis.[101] One group reported "... intracellular deposits of Aβ can occur early in the natural history of Aβ formation, and that they are associated with impaired behavior, underscore their potential role as therapeutic targets for disrupting the amyloid cascade, and for rescuing related functional impairments."[102]

In 2002, intracellular Aβ was proposed as the target for AD in drug discovery.[103] A plausible mechanism detailed the neuropathological processes leading to AD: Aβ42 internalization into neurons that involves the high-affinity interaction between Aβ42 and the α7 receptor leading to the internalization and intracellular accumulation of the Aβ42/α7 receptor complex. This alternative hypothesis can account for many of the well-known features of AD pathology, including specific cholinergic and cholinoceptive neuronal and synaptic loss, that affects cognitive and memory functions, the distribution, morphology, and composition of dense-core plaques, and their association with inflammation. More importantly, it provides the scientific rationale for targeting the mechanisms that lead to the intraneuronal Aβ42 as a novel strategy for AD drug discovery.

Another early report detected intracellular Aβ42 in neurons with the presumption that it accumulates from within the neuron. The accumulation of Aβ42 within neurons was proposed to represent an early pathological step in the events leading to AD.[104]

◄ despite the presence of fibrillar plaques in 100% of individuals with Down syndrome the fifth decade; (12) The *APOE4* genotype has numerous functional effects, rather than solely relating to reduced Aβ clearance, including links to enhanced inflammatory phenotypes. *Aβ,* amyloid-β; *APP,* amyloid precursor protein; *APOE4,* apolipoprotein E4; *FAD,* familial Alzheimer's disease. *Source: Used with permission from* Acta Neuropathol Commun *2014;2:135.*

In 2004, a modification to the amyloid cascade hypothesis calling into attention intraneuronal Aβ was proposed.[101] This hypothesis continues to incorporate similar elements of the amyloid cascade hypothesis such as FAD-related risk factors that eventually lead to accumulation of intraneuronal Aβ42 and Aβ40 levels. The major difference from the Inside-Out hypothesis lies in how amyloid plaques form.[105] The modified amyloid hypothesis retains the extracellular deposition process of amyloid plaque formation, while the Inside-Out hypothesis proposes that amyloid plaques are simply the lysed neurons. In 2005, another alternative hypothesis was presented on the basis that secreted Aβ peptides accumulate in the extracellular spaces of the CNS to form into toxic oligomers. These oligomers kill neurons and eventually form deposits of senile plaques that may never leave the membrane to exert their toxic effects within the neuron.[106]

Intraneuronal Pathological Consequences of Intraneuronal Aβ

It is unclear of pivotal reports linking intraneuronal Aβ to neuronal degeneration because intraneuronal Aβ was reported years ago when its presence was only considered as a source of the extracellular Aβ in the form of senile plaques. Possibly, one of the first publications that may have redirected the focus back on the significance of intraneuronal Aβ reported the link between intraneuronal Aβ and neuronal death with plaque formation.[92,97,98] Around the same time, another group reported the detection of intracellular deposits of Aβ42 in the PS1 transgenic mice without the presence of amyloid plaque formation showing that pathogenic role of the PS1 mutation was upstream of the amyloid cascade.[107] Also around the same time, another group detected Aβ42 in neurons strengthening the belief that intraneuronal Aβ can lead to the progression of the disease.[104]

There are two very important aspects of intracellular Aβ to report through the remainder of this book. The first is to discuss the pathological consequences of intraneuronal Aβ within the neuron (below), and the second is to discuss the pathological consequences of intracellular Aβ in relation to plaques (see Chapter 5), cognitive impairment (see Chapter 6), inflammation (see Chapter 7), and to the vascular system (see Chapter 8).

To begin, some of the reported intracellular consequences of Aβ are shown in Fig. 4.4.

Mouse models of AD show dramatic intraneuronal Aβ accumulation and neuronal cell death that precedes Aβ deposition.[108] A clear relationship between intraneuronal Aβ accumulation and neuronal loss is evident in transgenic mouse models.[109,110] The abnormalities and cognitive

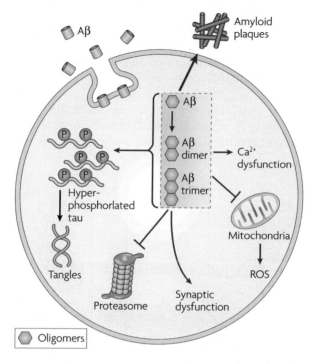

FIGURE 4.4   Pathological effects of intraneuronal Aβ. Aβ produced intracellularly or taken up from extracellular sources has various pathological effects on cell and organelle function. Intracellular Aβ can exist as a monomeric form that further aggregates into oligomers, and it may be any of these species that mediate pathological events in vivo, particularly within a dysfunctional neuron. Evidence suggests that intracellular Aβ may contribute to pathology by facilitating tau hyperphosphorylation, disrupting proteasome, and mitochondria function, and triggering calcium and synaptic dysfunction. *Aβ*, amyloid-β; *Ca*, calcium; *ROS*, reactive oxygen species. *Source: Used with permission from* Nat Rev Neurosci *2007;8:499−509.*

dysfunctions in several models of AD correlated with the appearance of intraneuronal Aβ, well before the appearance of plaques or tangles.[46,111] As presented in Fig. 4.4, the pathological effects of intraneuronal Aβ extend into many of the subcellular systems, which are detailed below.

On Intermediate Filaments

The intraneuronal cytoskeleton is composed of microtubules that are important in a variety of intracellular functions such as intracellular transport, movement of secretory vesicles, and organelles. Microtubule-associated protein 2 (MAP2) stabilizes the microtubules of the neuronal dendrites, while tau stabilizes the microtubules of the neuronal axon. Abnormal morphology and function of MAP2 and tau in AD neurons may contribute to neuronal death.[112,113]

Intraneuronal Aβ accumulated in dendrites and postsynaptic compartments that increased in age the Swedish mutant transgenic 2576 mouse model of AD (Fig. 4.5).[113] As the Aβ accumulated, the levels of MAP2 decreased without much effect on tau-1. The subcellular accumulation of Aβ42 was spatially associated with very early MAP2 pathology, which was observed previously.[114–118] A loss of MAP2 immunoreactivity was observed in areas of senile plaques in the Alzheimer's brain.[112] It was not clear if intraneuronal levels of Aβ were the cause of the abnormal MAP2 expression, but it was clear that

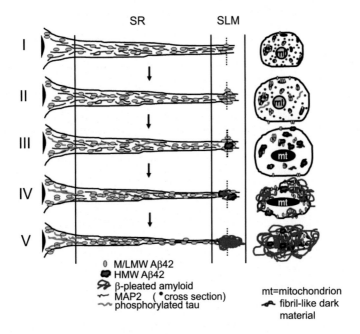

FIGURE 4.5 Schematic diagram of proposed sequence of Aβ42, MAP2, and tau alterations in a CA1 pyramidal cell apical dendrite with aging. At top (I), normal dendrite contains MAP2 associated with microtubules and M/LMW Aβ42 peptides. Cross section taken through the distal apical dendrite (SLM) is shown at right. (II) With aging, M/LMW Aβ42 peptides accumulate, which coincides with early reductions in MAP2, especially in the SLM. Later on in Tg2576 mice, Aβ42 M/LMW peptides co-localize with hyperphosphorylated tau in distal processes and synaptic compartments; this co-localization is more prominent in 36Tg mice (23). (III) Subsequently, Aβ42 HMW oligomers develop in the distal dendrite, which is associated with localized absence of MAP2. Concomitantly, Aβ42 M/LMW peptides further accumulate in more proximal regions of the dendrite. (IV) This is followed by Aβ fibril formation, especially in distal neurites of the SLM and (V) deposition of amyloid plaques in SLM. *Aβ42*, amyloid-β42; *HMW*, high molecular weight; *LMW*, low molecular weight; *MAP2*, microtubule-associated protein 2; *M/LMW*, monomer and low molecular weight; *mt*, mitochondrion; *SLM*, stratum lacunosum-moleculare; *Tg2576*, Swedish mutant transgenic mice. *Source: Used with permission from PLoS One 2013;8(1):e51965.*

abnormal MAP2 immunolabeling in the AD brain suggested focal areas of toxicity that appeared to be areas of neuronal lysis (dense-core plaques).

NFTs, another neuropathological feature in the AD brains, are composed of intracellular aggregates of the hyperphosphorylated microtubule-associated protein tau.[2] In normal healthy neurons, the axonal protein tau stabilizes microtubules that form the cytoskeleton of the cell by a process involving phosphorylation and dephosphorylation of the protein, and is therefore critical for normal neuronal activity in the brain. When tau is phosphorylated, it is unable to bind microtubules, and instead polymerizes with other tau molecules forming straight filaments that subsequently form paired helical filaments, which affect cellular geometry leading to a failure of neuronal transport and eventual cell death leaving behind a marker of the neuron, referred as a ghost tangle. It is not clear if these tangles are the cause or product of neuronal death since widespread neuronal loss can occur before their formation, though understanding the relationship between intracellular Aβ and tau is critical in AD research.[113]

In FAD-transgenic rats, the presence of intraneuronal Aβ induced the upregulation of the phosphorylated form of ERK2, and its enzymatic activity in the hippocampus independent of any changes in the activity or phosphorylation status of other putative tau kinases.[119] The increase in active phospho-ERK2 was accompanied by increased levels of tau phosphorylation suggesting that in the absence of plaques, intraneuronal accumulation of Aβ peptide correlated with the initial steps in the tau-phosphorylation cascade, and alterations in ERK2 signaling in male rats.

The pathological accumulation of intraneuronal Aβ42 within postsynaptic compartments of dendrites in CA1 neurons of the hippocampus correlated with early pathological redistribution and hyperphosphorylation of tau in the Swedish mutant transgenic mice (Fig. 4.5).[120] In triple-transgenic mice, aberrant accumulation of Aβ and phosphorylation of tau occurred early and prominently in distal dendrites and postsynaptic compartments as paired helical filaments.[121] At older ages of the Swedish mutant transgenic 2576 mice, hyperphosphorylated tau co-localized with accumulating Aβ42 peptides within dystrophic neurites around plaques.[113] A direct interaction between tau protein and Aβ with the results of tau aggregation and TPK II-mediated phosphorylation of this protein has been reported to occur in vitro.[122]

Intraneuronal Aβ may negatively affect tau indirectly through other mechanisms. In vitro findings using neuroblastoma and ex vivo synaptosomes demonstrated increased phosphorylation of tau (Ser-202, Thr-181, and Thr-231) following Aβ peptide binding to the α7 receptor, which thereby represents a receptor potentially involved in amyloid-induced tau pathology.[123] Aβ-induced reduction of NMDA-dependent LTP was linked to increased phosphorylation of tau and glycogen synthase kinase 3 (GSK3) activation in hippocampal slices.[124−126] These data could point to a role of NMDA receptors in Aβ-induced changes in tau, possibly via GSK3β.[123] Increased expression of GSK has been observed in two different models of Aβ-induced tau pathology, whereas inhibition of GSK in these models resulted in reduced tau pathology.[123,127,128]

Intraneuronal aggregates of Aβ42, as well as that involving tau phosphorylation by GSK3 promoted by Aβ, could explain how Aβ peptide promotes the formation of tau aberrant aggregates in tau-Aβ transgenic mice.[129] Besides the tau aggregation induced by Aβ peptide, another consequence of the interaction between Aβ and tau protein is the decrease in Aβ aggregation. The disassembling action shown by tau protein on the 11 residues Aβ 25−35 suggests the possibility that tau protein may have a protective role preventing Aβ from acquiring the cytotoxic, aggregated form.[130]

It is not clear if normal physiological levels of intraneuronal Aβ normally phosphorylate tau, or at what point levels of intraneuronal Aβ42 become pathological to cellular processes; regardless, intraneuronal Aβ can also hyperphosphorylate tau, thereby further highlighting the importance of intracellular amyloid in AD.

On the Endolysosomal Pathway

The movement of molecules in and out of the cell requires orchestration of an elaborate system of cellular components, which are also responsible for the sorting, digestion, and storage of material. Fig. 4.6 shows general examples of some of these sorting cellular components such as endosomes, lysosomes, and multivesicular bodies (MVBs) albeit in fungi.[131]

Aβ42 is present in normal neurons of young and nondemented age-matched control human brains; however, the amounts of intraneuronal Aβ differ considerably in the neurons of AD brains. Initial reports of intraneuronal Aβ in AD brains show the accumulation of Aβ42 in the perikaryon of pyramidal neurons as discrete granules that appeared cathepsin D-positive suggesting they represent lysosomes or lysosome-derived structures supporting an important role for lysosomes in AD pathogenesis.[92,98] Significant intraneuronal Aβ42 was frequently co-localized with the lysosomal marker, LAMP-1, in the murine hippocampus and frontal cortex of PDAPP mouse model of AD.[132] In addition to lysosomes, intraneuronal Aβ was observed in abnormal endosomes, MVBs, and within pre- and postsynaptic compartments.[133−136]

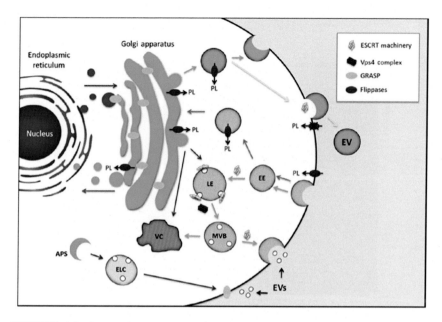

FIGURE 4.6 Potential participation of components of the endosomal sorting complex required for transport (ESCRT) machinery, Golgi reassembly stacking protein (GRASP) and flippases in the biogenesis of fungal EVs. The similarities between EVs produced by fungi and mammalian exosomes suggest that ESCRT machinery is required for formation of the fungal compartments (*green arrows*). Maturation of the late endosome (LE) is accompanied by membrane invagination, giving origin to small intraluminal vesicles and MVBs. The ESCRT machinery is recycled through the activity of the Vps4 protein complex. MVB may be directed to VC degradation pathways, but also to fusion with the plasma membrane, releasing exosomes to the extracellular milieu now receiving the name exosomes. GRASP, a regulator of unconventional secretion by mechanisms that are putatively linked to EV release was first identified as a structural component of the Golgi cisternae. Alternative roles included tethering activity for endosomal or lysosomal compartments and/or regulation of autophagy-related mechanisms (*blue arrows*). GRASP may also localize to the plasma membrane, mediating the release of exosomes to the extracellular space. Finally, GRASP can also participate in docking or fusion events involving vesicles originating at the Golgi, thus facilitating anterograde transport through the early secretory pathway (*brown arrow*). Flippases, an ATP-dependent enzyme responsible for membrane lipid bilayer asymmetry, are involved in vesicle biogenesis through phospholipid translocation across the lipid bilayers. These enzymes can regulate endocytosis at the plasma membrane level (*orange arrows*) and also drive the formation of exocytic vesicles (*light blue arrows*). Elevated calcium levels and oxidative stress can lead to a loss of flippase functions leading to an abnormal distribution of phospholipids in AD. Flippases can also participate in protein trafficking between the trans-Golgi network and endosomal compartment or between the trans-Golgi network and vacuoles (*purple arrows*). It has been also proposed that flippases may regulate the retrograde transport pathway from the Golgi apparatus to the ER (*pink arrow*), as well as vesicle budding at the plasma membrane level (*yellow arrow*). The possibility that cellular pathways regulated by endosomal proteins, GRASPs, and flippases are interconnected cannot be ruled out, as previously described for other unconventional secretory pathways. Most of the mechanisms proposed here have been implicated with the physiology of yeast cells, although they also participate in pathways required for molecular degradation and/or export in other eukaryotes. *APS*, autophagosome; *EE*, early endosome; *ELC*, endosomal/lysosomal compartment; *LE*, late endosome; *MVB*, multivesicular bodies; *PL*, phospholipid; *VC*, vacuole. *Source: Used with permission from* Int J Mol Sci 2013;*14*:9581−9603.

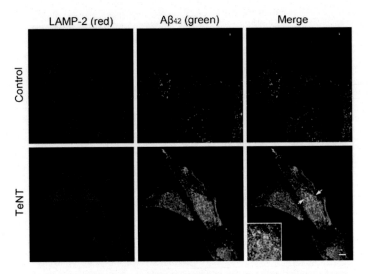

FIGURE 4.7   Double immunostaining for Aβ42 (green fluorescence) and LAMP2 (red fluorescence) in retinoic acid differentiated SH-SY5Y cells cultured under normal conditions or exposed to exocytosis inhibitor TeNT (5 nM) for 24 h. Both Aβ42 and LAMP-2 specific fluorescence are increased in tetanus toxin exposed cells suggesting the increased amount of intracellular Aβ42, as well as the size of the lysosomal compartment, and the co-localization of Aβ42 with LAMP-2 positive structures (*arrow* and corresponding inset) is increased after TeNT treatment. Scale bar, 5 μm. *Aβ*, amyloid-β; *LAMP2*, lysosomal/late endosomal marker; *TeNT*, tetanus toxin. *Source: Used with permission from* Transl Neurodegener *2012;1:19.*

Under normal conditions, the majority of intraneuronal Aβ was localized in extra lysosomes suggesting that cells are able to perform a rapid proteolytic digestion of this peptide (Fig. 4.7).[137] However, under oxidative stress, intraneuronal Aβ is significantly increased in the lysosomes through activation of macroautophagy.[137,138]

Aβ42 can be imported into the cell through the endosomal/lysosomal system and, thus, may be less exposed to the Aβ-degrading enzymes, neprilysin, and insulin-degrading enzyme.[139,140] The presence of Aβ42 within small organelles such as lysosomes would favor the aggregation of Aβ42 because of the acidic pH characteristic that favors oligomerization.[141,142] Aβ42 aggregates more rapidly, and is more resistant to degradation than Aβ40, and preferential internalization of Aβ42 from medium has also been reported in CA1 neurons in rat organotypic slice cultures.[143,144]

Extracellular Aβ can internalize in neurons by receptor-mediated endocytosis. In PC12 cells overexpressing the low-density lipoprotein-related protein minireceptor (mLPR2), Aβ42 binds to the receptor, and is internalized directly and through ApoE. The endocytosed Aβ42 is not completely degraded and accumulates in intraneuronal lysosomes.[137] Significant amounts of intraneuronal Aβ42 in 3-month-old PDAPP mice, most of which was localized inside lysosomes, were also observed. *APOE4*-transgenic mice housed in an enriched environment showed increased levels of oligomerization and deposition of Aβ peptides in hippocampal neurons compared to *APOE3*-transgenic mice housed under the same conditions.[145] Inhibition of the Aβ-degrading enzyme neprilysin in *APOE3* and *APOE4* mice resulted in an *APOE4* isoform-specific degeneration of hippocampal CA1 neurons, which was accompanied by intracellular accumulation of Aβ and ApoE and lysosomal activation.[146] In agreement with this lysosomal pathology, altered endocytic pathway activity was one of the earliest known intraneuronal changes occurring in sporadic AD, and the *APOE4* genotype promoted early-stage endocytic pathway activation.[133,147,148]

Receptor-mediated endocytosis of extracellular Aβ was also observed in α7 receptor-transfected human neuroblastoma cells.[149] The rate and extent of Aβ42 internalization was directly related to the level of α7 receptor protein since the rate of Aβ42 accumulation was much lower in untransfected cells that express much lower levels of this receptor. Also, internalization of Aβ42 was selectively blocked by α-bungarotoxin, an α7 receptor antagonist, and by phenylarsine oxide, an inhibitor of endocytosis. As in neurons of AD brains, the α7 receptor in transfected cells was precisely co-localized with Aβ42 in prominent intracellular aggregates in lysosomal-like structures in the neuronal perikaryon.

Lysosome dysfunction was linked to synaptic pathology in the AD brain, and inhibition of lysosome acidification with chloroquine resulted in synaptic dysfunction and synaptic loss.[150-154] To that point, early abnormalities in the endosomal-lysosomal system (eg, enlarged endosomes) preceded extracellular amyloid deposition (ie, amyloid plaque) in sporadic AD and Down syndrome.[133,155] However, it was suggested that endosomal pathology contributed significantly to neuronal Aβ production and accumulation.[133]

Although the lysosomal system can degrade Aβ42 in cells, the ubiquitin-proteasome pathway may also degrade Aβ42 in neurons as well.[92,136,156-158] Because the ubiquitin-proteasome system may be affected in AD brain and contribute to neuronal degeneration, activation of Aβ42 degradation by proteasome would attenuate intracellular Aβ42 pathogenesis.[159,160]

The ubiquitin-proteasome activity decreases with age in the brains of Tg2576 mice, while the Aβ42 levels increased.[161] In cultured neuronal cells, an extracellular treatment of Aβ markedly decreased the proteasome activity, and extracellular treated Aβ peptides were found in the cytoplasmic compartment. Aberrant accumulation of Aβ42 in MVBs, which are considered to be late endosomes formed from early endosomes by membrane invaginations that generate inner vesicles with lower pH (~5.5), may impair MVB sorting within the neuron by inhibiting the ubiquitin-proteasome system.[162,163] Aβ42 is localized predominantly to MVBs of neurons in normal mouse, rat, and human brain.[134,135,163,164] In transgenic mice and human AD, intraneuronal Aβ42 increased with age, and Aβ42 accumulated in MVBs within presynaptic and especially postsynaptic compartments. This accumulation was associated with abnormal synaptic morphology before Aβ plaque pathology suggesting that intracellular accumulation of Aβ is an early pathological event, and plays a crucial role in AD.

APP mutant neurons demonstrated impaired inactivation, degradation, and ubiquitination of epidermal growth factor receptor (EGFR) when stimulated with EGF.[163] EGFR degradation was dependent on translocation from MVB outer to inner membranes, which is regulated by the ubiquitin-proteasome system. These data suggest a mechanism whereby Aβ accumulation in neurons impairs the MVB sorting pathway via the proteasome activity in AD.[161,163]

In the majority of sporadic AD cases, since there is no consistent overproduction of Aβ, deficits in its degradation could lie behind the pathogenesis of the disease. Regardless of how Aβ accumulated in neurons, its accumulation can compromise normal neuronal function in AD.[137]

On Chaperone Proteins

Aβ can co-localize with several chaperone proteins in a transgenic *C. elegans* model designed to express a chimeric signal peptide/Aβ42 minigene to route intracellular Aβ into the secretory pathway and intracellular Aβ.[165–167] One of the major functions of chaperones is to assist in the covalent folding or unfolding, and the assembly or disassembly of other proteins. These data suggest that these proteins (two members of the HSP70 family, three α-B-crystallin-related small heat shock proteins [HSP-16s], and a putative ortholog of a mammalian small glutamine-rich tetratricopeptide repeat-containing protein proposed to regulate HSP70 function) may play an early role in Aβ metabolism.

In this model, Aβ is recognized as an abnormal protein, and actively rerouted from the secretory pathway to an alternative compartment for refolding or degradation.[167] This metabolism of Aβ may parallel that of prion protein (PrP) expressed in transfected cells, which is apparently continually subjected to ER quality control and retrograde transport, and where proteasome inhibitor treatment leads to cytoplasmic accumulation of PrP co-localized with HSP70.[168]

On Synapses

Significant synaptic loss ranging from 20% to 50% occurs in the AD brain, and the decline in cognition correlates best with synaptic loss and not extracellular amyloid plaque counts, implying that synaptic perturbations cause AD and precede amyloid plaque deposition.[31,121,169–173] In the triple-transgenic mice, synaptic dysfunction, including LTP deficits, manifests in an age-related manner but before plaque and tangle pathology.[46] Synaptic loss was proposed in 1991 as one of the first neuropathological events leading to AD, which can effect changes in cognition before the appearance of structural changes.[173,174]

Decreases in the numbers of presynaptic terminals, synaptic vesicles, and synaptic protein markers have also been reported.[3] The majority of excitatory synapses in the brain are made on the heads of dendritic spines.[175] Initially, synapse degeneration begins at the level of dendritic spines, the loci of memory-initiating mechanisms.[176–178] In the AD and transgenic mouse AD models, significant decreases occur in spine density, and in molecules involved in spine signaling and control of filamentous actin.[115,172,179–183]

Toxic effects of extracellular Aβ on synapses have been known for many years (Fig. 4.8). Although this model is based on extracellular Aβ secreted from the neurons, it clearly conveys events that begin with increased accumulation of extracellular Aβ oligomers in the synapses.[121,184,185] Aβ oligomers then promote endocytosis of NMDA and α-amino-3-hydroxy-5-methyl-4-isoxazolepropionic acid (AMPA) receptors, leading to a reduction in dendritic spines and reduced LTP, which is the basis of dysfunctional learning, memory, and complex information processing. Acceleration of this process could lead to synaptic dysfunction, spine loss, and (potentially) synaptic loss leading to cognitive decline and AD (Fig. 4.8).[121,185–187]

A revised synaptic hypothesis model considered intracellular Aβ as the cause and predictor of synaptic decline leading to cognitive impairment in the synapse rather than extracellularly, and that continuous overproduction of Aβ can oligomerize forming toxic aggregates, which subsequently reduce the number of synapses and impair synaptic plasticity (Fig. 4.9).[50,106,120,164,188–196] Intraneuronal accumulation of different Aβ peptides, together with overall accumulation of fibrillar and oligomeric species, coincided with 30% CA1 neuron loss, 18% hippocampus atrophy, and drastic reduction of synaptic plasticity.[38]

A relationship between synaptic activity and levels of intracellular Aβ showed that synaptic activity decreased intracellular Aβ in primary neuronal culture. This was also observed in 4-month-old Tg19959 mice, which overexpress human APP carrying the Swedish (K670N/M671L) and Indiana (V717F) mutations likely by enhancing Aβ degradation.[186,197] Similarly, synaptic activation reduced levels of intraneuronal Aβ and

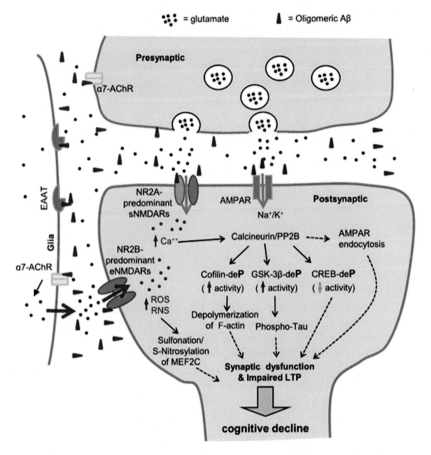

FIGURE 4.8   Schematic diagram outlining mechanisms of oligomeric Aβ-induced synaptic dysfunction. At pathological concentrations, Aβ oligomers may interact with multiple astrocytic, microglial, and neuronal synaptic proteins, including α7-AChRs and NMDARs, triggering a series of toxic synaptic events. These events include aberrant activation of NMDARs (especially NR2B-containing extrasynaptic NMDARs), elevated neuronal calcium influx, calcium-dependent activation of calcineurin/PP2B and its downstream signal transduction pathways, involving cofilin, GSK-3β, CREB, and MEF2. This results in aberrant redox reactions and severing/depolymerizing F-actin, tau-hyperphosphorylation, endocytosis of AMPARs, and eventually leads to synaptic dysfunction and cognitive impairment. *Aβ*, amyloid-β; *α7-AChRs*, α7 nicotinic acetylcholine receptors; *AMPARs*, α-amino-3-hydroxy-5-methyl-4-isoxazolepropionic acid receptor; *CREB*, cyclic AMP response element-binding protein; *GSK-3β*, glycogen synthase kinase 3β; *MEF2*, transcription factor myocyte enhancer factor 2; *NMDARs*, N-methyl-ᴅ-aspartate receptors. *Source: Used with permission from Mol Neurodegener 2014;9:48.*

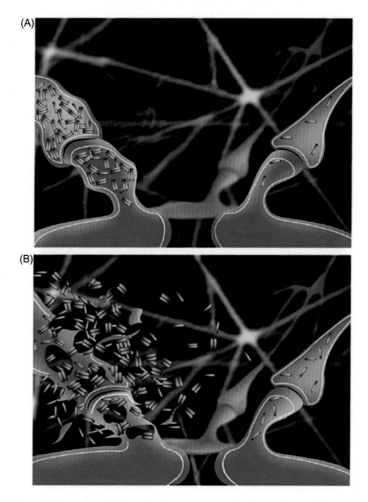

FIGURE 4.9   (A) Accumulation of intraneuronal Aβ42 at early stages of AD occurs progressively in a synapse (left) and is associated with pathological alterations compared to a normal synapse (right). (B) Intraneuronal Aβ42 is released in the extracellular space following degeneration of the synapse (left). Release of intraneuronal Aβ42 into the extracellular space may contribute to the toxic spread of Aβ pathology to a nearby synapse (right). *Aβ42*, amyloid-β42. *Source: Used with permission from* Front Aging Neurosci 2010;2(13):1–5.

protected against Aβ-related synaptic changes. It was also demonstrated that synaptic activity promoted APP transport to synapses, and that the Aβ-degrading protease neprilysin is involved in the activity-induced reduction of intraneuronal Aβ42.[186]

These two models of synaptic degeneration begin differently (extracellular accumulation vs intracellular accumulation) but end similarly with extracellular toxic Aβ. However, the classical belief that extracellular Aβ is toxic continues to be challenged. The possible toxic effect of extracellular soluble Aβ species is likely to be negligible because other groups of neurons like dentate gyrus granule cells, which do not accumulate intraneuronal Aβ, were unaffected in the APP/PSI KI bigenic mice.[38] Intra-axonal oligomeric Aβ42, but not oligomeric Aβ40 or extracellular oligomeric Aβ42, inhibited synaptic transmission.[198] Similar nM levels of Aβ disrupt hippocampal LTP.[194,199]

On Mitochondria

Mitochondria are dynamic adenosine triphosphate (ATP)-generating organelles that contribute to a myriad of cellular functions including bioenergetics processes, intracellular calcium regulation, alteration of reduction-oxidation potential of cells, free radical scavenging, and activation of caspase-mediated cell death.[200] In the neurons, mitochondria are located in areas that have the highest demands for ATP, such as in the synapses. Mitochondrial functions can be negatively affected by intracellular Aβ as observed in AD and Down syndrome brains.[123,201−203]

Mitochondria serve as direct targets of neuronal toxicity in which Aβ associates with the outer mitochondrial membrane, intermembrane space, inner mitochondrial membrane, and the matrix.[121] Aβ interacts with cyclophilin D, a protein component of the membrane permeability transition pore (MPTP). The interaction of cyclophilin D with Aβ causes functional modification of this protein leading to MPTP opening. Aβ also binds with another mitochondrial protein, Aβ alcohol dehydrogenase (ABAD) to distort the enzyme's structure, rendering it inactive. The ABAD−Aβ complex, which is upregulated in neurons from AD patients, can induce oxidant stress and mitochondrial dysfunction. This causes an increase in reactive oxygen species and oxidative stress leading to initiation of apoptosis.

Intracellular Aβ progressively accumulates in the mitochondria of neurons in the AD brain and as well as in the brains of transgenic mouse models of AD.[200] This accumulation of Aβ in the mitochondria in the transgenic mice arises before the detection of extracellular deposition, and as early as 4−5 months while increasing with age. Intra-mitochondrial Aβ can inflict a myriad of pathologies that include oxidative stress, abnormal dynamics, deficits in bioenergetics, impaired biogenesis, etc. Although it is unclear if the intra-mitochondrial Aβ is generated in situ, or imported from within the neuron, the former possibility is unlikely since it is not known if γ-secretase is associated with mitochondria. However, the latter seems likely because when extracellular Aβ was internalized into the cells, it was subsequently co-localized with mitochondrial markers suggesting a specific uptake mechanism for intracellular Aβ that may involve the translocase of the outer membrane complex.

Intracellular Aβ can also induce permeabilization of mitochondrial membrane and cytochrome C release indirectly through the activation of the p53 protein, which inhibits the cell cycle and induces apoptosis.[156,204–206] The expression of p53 was also associated with synaptic degeneration as well as mitochondrial dysfunction.[207]

Yeast cells have been used to study the effects of intracellular Aβ peptides.[208] Cells that constitutively produce native Aβ directed to the secretory pathway exhibited a lower growth rate, lower biomass yield, lower respiratory rate, and increased oxidative stress; hallmarks of mitochondrial dysfunction and ubiquitin-proteasome system dysfunction.

On Calcium Signaling

Synaptic activity is essential for neuronal survival, and is regulated by the entry of appropriate amounts of calcium through synaptic receptors such as the α7 and NMDA receptors.[209] Properly controlled homeostasis of calcium signaling not only supports normal brain physiology, but also maintains neuronal integrity and long-term cell survival through apoptosis suppression and survival promotion.[121]

Aβ can act through multiple targets including calcium channels and various receptors in membranes. Aβ directly interacts with voltage-dependent calcium channels and TRPC to produce a transient increase in calcium necessary for synaptic plasticity and neuronal survival.[121] Aβ enhances transmitter release by transient increase of glutamate release from the presynaptic terminal that results from brief periods of high-frequency stimulation with calcium build-up within the terminal to trigger short-term synaptic plasticity.[210] Aβ peptides bind to the α7 receptor with high affinity, and low pM levels of Aβ activated these receptors at presynaptic nerve endings of synaptosomes producing an influx of calcium.[211–213]

Dysregulation of intracellular calcium has been implicated in AD, which may be indirectly due to the toxic levels (exceeding physiological levels) of intraneuronal Aβ.[214] Such toxic levels of intraneuronal Aβ could deregulate calcium signaling mechanism by excessive accumulation of calcium in the cytoplasm and cytoplasmic organelles such as mitochondria, leading to compromised synaptic function, and eventual neuronal degeneration.[121,215]

Many calcium-dependent secondary messaging systems are impaired in AD such as cyclic adenosine monophosphate (cAMP) signaling. Levels of the activated (ie, phosphorylated) form of cAMP response element-binding protein (CREB) are reduced in AD compared to that of an age-matched, healthy control group.[216] Calcium signaling to the cell nucleus is the key inducer of CREB phosphorylation on its activator site serine 133.[217] Experiments in aged neurons show altered calcium signaling at the level of either calcium signal generation and/or calcium signal propagation.[218] These studies indicate a critical role of calcium in Aβ-induced synaptic activity and memory formation by regulating specific signal transduction pathways.

Therefore, the effects of Aβ can lead to an excessive intraneuronal calcium load through the overactivation of these receptors, or through disrupting the regulation of calcium homeostasis resulting in neuronal degeneration and death.[219,220]

On Apoptosis

The incidence of neuronal apoptosis is increased in AD, and is linked to intraneuronal Aβ and to autoimmunity.[221-228] The selective accumulations of Aβ42 in neurons undergo apoptosis in vitro suggesting that intracellular Aβ may lead to an altered cellular metabolism to promote neurodegeneration.[226] When primary guinea pig neurons were treated with hydrogen peroxide, an inducer for genomic DNA damage, p53 expression was detected along with the translocation of cytosolic Aβ42 into the nucleus.[155] The p53 expression was only detected in markedly Aβ42-accumulating and putatively degenerating neurons in transgenic mice.

Intracellular Aβ42, but not extracellular Aβ, induced neuronal apoptosis in vitro.[229] Also, the accumulation of Aβ42 in neurons occurred without the detection of Aβ42 in the extracellular spaces in the aged mice carrying the PS1 gene.[230] Similarly, it was suggested that a mechanism of intracellular Aβ-induced apoptosis may be through its effect on the p53 promoter, which is also upregulated in AD and Down syndrome brains.[225,231-235] The p53-dependent apoptosis of primary neurons by Aβ42 injections in the cytosol has been reported.[236]

Mutations in the PS1 gene, which account for about 50% of patients with FAD, may also predispose neurons to apoptotic death.[237-239] Expression of mutant PS1 in PC12 cells increased their vulnerability to apoptosis induced by Aβ and trophic factor withdrawal.[240,241] Overexpression of calbindin in neural cells counteracted the pro-apoptotic actions of PS1 gene mutations by a mechanism involving stabilization of intracellular calcium levels, suppression of oxidative stress, and preservation of mitochondrial function.[227]

The AICD also has cytotoxic effects on neuronal cells, and participates in the regulation of gene activation.[242] Detailed analysis demonstrated that the AICD also interacts with p53 and enhances its transcription and pro-apoptotic functions.[243] Once the AICD is enzymatically cut from the full-length APP protein, it is translocated into the nucleus by an adaptor protein Fe65.[243,244] The inhibition of the enzymatic activity (eg, β-secretase) led to significant reductions in p53-mediated transcriptional activation, as well as resistance of apoptotic stimuli in FAD-associated, APP mutant-expressing cells.[245] It is likely that intracellular Aβ and/or nuclear AICD may induce neuronal apoptosis at least in part through the p53-dependent pro-apoptotic pathway.[243]

In addition to the toxic influences of intraneuronal Aβ on neurons, apoptotic neurons were also associated with immunoglobulin-positive neurons in the AD brains.[228] Apoptotic neurons were defined by caspase-3 expression, a marker of irreversible apoptotic cell death, and by their morphological features (eg, cell atrophy, degenerating processes, condensed, and pyknotic, nuclear chromatin). Nearly all (>98%) of the apoptotic neurons in the AD brain were also positive for immunoglobulin immunoreactivity suggesting an autoimmune-dependent apoptotic pathway of neuronal degeneration, which may also be related to intraneuronal accumulations of Aβ.

The microinjection of intracellular Aβ42 into neurons induced dramatic cell death mediated through the activation of apoptotic-related markers: p53, Bax, and caspase-6.[236,246] Also, intraneuronal Aβ42 was linked to neuronal apoptosis in transgenic mice.[247] Transgenic mice expressing Aβ42 in neurons showed extensive neurodegeneration, which presumably occurred through an apoptotic pathway.[248] Intracellular Aβ deposits correlated with cell damage and apoptotic cell death, as well as synaptic pathology in the AD brain.[249] Treatment of rat cortical neurons with a functional γ-secretase inhibitor strongly reduced the production of intracellular Aβ42, and improved cell survival.[229] Intracellular accumulation of Aβ42 increases the vulnerability of CA1 pyramidal neurons in AD.[250]

Morphological and Accumulating Levels Over Time

As previously noted, the Aβ immunoreactivity was granular in nature in the brains of young monkeys as young as 4 years old.[251] This diffuse pattern of immunoreactivity indicated that the intracellular Aβ in young monkeys may represent Aβ generated biologically in neurons, rather than accumulated Aβ that is typically found in neurons of aged animals. In the brains of aged monkeys (24–36 years old), Aβ immunoreactivity in some cortical neurons tended to cluster together in clumps. This clustering of Aβ in neurons is consistent with previous hypotheses that Aβ accumulates in lipid rafts, and that clustering of GM1 ganglioside within lipid raft-like membranes may facilitate the aggregation of Aβ (fibrils) in these rafts.[252] Increased levels of Aβ42 occurred over time, while there was no change in the Aβ40 species.[253]

No intraneuronal Aβ aggregates were observed in the neurons in the 1-month-old APPSwe transgenic mouse suggesting that intraneuronal Aβ aggregates increase with age, and predate plaque formation also in APP transgenic mice that do not contain the Arctic APP mutation.[253]

In contrast, at 7 months of age, aggregates of intraneuronal Aβ were detected that were similar to that in APPArcSwe transgenic mice. However, the intensity and frequency of the intraneuronal Aβ aggregates was reduced in APPSwe transgenic mice compared to that of APPArcSwe transgenic mice of a similar age; neither had developed extracellular plaque deposition. At 9 months of age, grain-like intraneuronal Aβ aggregates were frequent and widespread in the cerebral cortex, subiculum, and the CA1 region of hippocampus in APPArcSwe transgenic mice with modest amyloid plaque deposition.

In non-AD, age-matched control human brain tissues, punctate, Aβ42-positive endosomal granules were observed in pyramidal neurons, which was in contrast to the excessive accumulations of Aβ42 in some of the dendrites and perikaryon of AD pyramidal neurons (Fig. 4.10).[92,254] Similarly, neurons from neurologically normal controls (ages 3 months

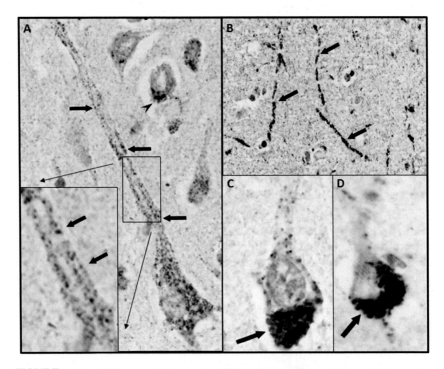

FIGURE 4.10   (A) Presence of punctate Aβ42-positive granules within and along representative non-AD, age-matched control pyramidal neurons (*arrows*). Note a nearby pyramidal neuron with minimal areas of Aβ42 accumulation (*arrowhead*). (B–D) Presence of Aβ42-positive material within AD distended pyramidal neuronal dendrites (B, *arrows*) and in the perikaryon (C, D, *arrows*) using routine IHC techniques without formic acid pretreatment as previously reported. *Aβ42*, amyloid-β42. *Source: Used with permission from Curr Pharm Design 2006;12(6):677–684.*

to 79 years) showed intraneuronal Aβ42 staining that appeared to increase in relation to the subject's age at death, and in AD brains, Aβ was also present in many swollen neuronal processes.[104,255]

Age-related intraneuronal Aβ42 was observed predominately within dystrophic neurites and synaptic compartments rather than in the cell bodies of the transgenic 2576 mice.[135,164] Another report showed oligomers of Aβ42 aggregates along microtubules of neuronal processes in cultured transgenic 2576 neurons, as well as in transgenic 2576 mouse and human AD brain neurons.[164,254]

In addition to reporting changing amounts of intraneuronal Aβ, the Aβ42 levels and the Aβ42/Aβ40 ratio were significantly higher in AD cases as compared with controls.[247]

SUMMARY

Generally, three scenarios exist for a pathological role of Aβ in neuronal death leading to AD:

1. Intraneuronal Aβ is secreted into the extracellular spaces where it accumulates over time as toxic aggregates becoming senile plaques, which in turn kill nearby neurons (the amyloid cascade hypothesis).
2. Intraneuronal Aβ accumulates either from within the neuron due to overproduction (FAD-dependent) or from uptake within the brain or from peripheral sources, which aggregate in synapse or terminal processes over time leading to the degeneration of neuronal processes and extracellular plaque formation leading to neuronal death.
3. Intraneuronal Aβ accumulates primarily from uptake within the brain or from peripheral sources that aggregate in the perikaryon of the neuron (ie, in lysosomes), which over time lead to neuronal death by lysis forming the amyloid plaque. Cell lysis then triggers additional neuronal death independent of amyloid (see Chapter 9).

Since scenario number 1 continues to fail in the clinic, the viability of scenarios 2 and 3 remain to be tested. If Aβ is going to continue to be the focus of AD research, reducing intraneuronal levels must be the target, possibly by inhibiting uptake through one of the Aβ-binding receptors, and/or prohibiting peripheral Aβ from gaining unregulated access into the brain.[97,103,255]

References

1. Hardy JA, Higgins GA. Alzheimer's disease: the amyloid cascade hypothesis. *Science* 1992;**256**(5054):184–5.
2. Morris GP, Clark IA, Visse B. Review. Inconsistencies and controversies surrounding the amyloid hypothesis of Alzheimer's disease. *Acta Neuropathol Commun* 2014;**2**:135.

3. Xiaqin S, Yan Z. Amyloid Hypothesis and Alzheimer's Disease. In: Chang R C-C, editor. *Advanced Understanding of Neurodegenerative Diseases*. InTech; 2011. ISBN: 978-953-307-529-7, Available from: < http://www.intechopen.com/books/advanced-understanding-of-neurodegenerativediseases/amyloid-hypothesis-and-alzheimer-s-disease>.
4. Hardy J, Allsop D. Review. Amyloid deposition as the central event in the aetiology of Alzheimer's disease. *Trends Pharmacol Sci* 1991;**12**(10):383−8.
5. St George-Hyslop PH, Petit A. Review. Molecular biology and genetics of Alzheimer's disease. *C R Biol* 2005;**328**:119−30.
6. Gandy S, Petanceska S. Regulation of Alzheimer β-amyloid precursor trafficking and metabolism. *Biochim Biophys Acta* 2000;**1502**:44−52.
7. Borchelt DR, Thinakaran G, Eckman CB, et al. Familial Alzheimer's disease-linked presenilin 1 variants elevate Aβ1- 42/1-40 ratio in vitro and in vivo. *Neuron* 1996;**17**:1005−13.
8. Fraser PE, Yang DS, Yu G, et al. Presenilin structure, function and role in Alzheimer disease. *Biochim Biophys Acta* 2000;**1502**:1−15.
9. Czech C, Tremp G, Pradier L. Presenilins and Alzheimer's disease: biological functions and pathogenic mechanisms. *Prog Neurobiol* 2000;**60**:363−84.
10. Guerreiro RJ, Gustafson DR, Hardy J. Review. The genetic architecture of Alzheimer's disease: beyond APP, PSENs and APOE. *Neurobiol Aging* 2012;**33**(3):437−56.
11. Kang J, Muller-Hill B. Differential splicing of Alzheimer's disease amyloid A4 precursor RNA in rat tissue: PreA4695 mRNA is predominantly produced in rat and human brain. *BBRC* 1990;**166**:1192−200.
12. Goate A, Chartier-Harlin MC, Mullan M, Brown J, Crawford F, Fidani L, et al. Segregation of a missense mutation in the amyloid precursor protein gene with familial Alzheimer's disease. *Nature* 1991;**349**:704−6.
13. Goldgaber D, Lerman MI, McBride OW, Saffiotti U, Gajdusek DC. Characterization and chromosomal localization of a cDNA encoding brain amyloid of Alzheimer's disease. *Science* 1987;**235**(4791):877−80.
14. Chartier-Harlin MC, Crawford F, Hamandi K, Mullan M, Goate A, Hardy J, et al. Screening for the β-amyloid precursor protein mutation (APP717: Val—Ile) in extended pedigrees with early onset Alzheimer's disease. *Neurosci Lett* 1991;**129**:134−5.
15. Chartier-Harlin MC, Crawford F, Houlden H, Warren A, Hughes D, Fidani L, et al. Early-onset Alzheimer's disease caused by mutations at codon 717 of the β-amyloid precursor protein gene. *Nature* 1991;**353**:844−6.
16. Haass C, Lemere CA, Capell A, Citron M, Seubert P, Schenk D, et al. The Swedish mutation causes early-onset Alzheimer's disease by β-secretase cleavage within the secretory pathway. *Nat Med* 1995;**1**:1291−6.
17. Levy E, Carman MD, Fernandez-Madrid IJ, Power MD, Lieberburg I, van Duinen SG, et al. Mutation of the Alzheimer's disease amyloid gene in hereditary cerebral hemorrhage, Dutch type. *Science* 1990;**248**:1124−6.
18. Wisniewski T, Ghiso J, Frangione B. Review. Biology of Aβ amyloid in Alzheimer's disease. *Neurobiol Dis* 1997;**4**:313−28.
19. Haass C, Schlossmacher MG, Hung AY, Vigo-Pelfrey C, Mellon A, Ostaszewski BL, et al. Amyloid β-peptide is produced by cultured cells during normal metabolism. *Nature* 1992;**359**:322−5.
20. Nilsberth C, Westlind-Danielsson A, Eckman CB, Condron MM, Axelman K, Forsell C, et al. The 'Arctic' APP mutation (E693G) causes Alzheimer's disease by enhanced Aβ protofibril formation. *Nat Neurosci* 2001;**4**:887−93.
21. Steiner H, Capell A, Leimer U, Haass C. Review. Genes and mechanisms involved in β-amyloid generation and Alzheimer's disease. *Eur Arch Psychiatry Clin Neurosci* 1999;**249**(6):266−70.

INTRACELLULAR CONSEQUENCES OF AMYLOID IN ALZHEIMER'S DISEASE

22. Guo Q, Fu W, Sopher BL, Miller MW, Ware CB, Martin GM, et al. Increased vulnerability of hippocampal neurons to excitotoxic necrosis in presenilin-1 mutant knock-in mice. *Nat Med* 1999;**5**:101−6.

23. Jankowsky JL, Fadale DJ, Anderson J, Xu GM, Gonzales V, Jenkins NA, et al. Mutant presenilins specifically elevate the levels of the 42 residue β-amyloid peptide in vivo: evidence for augmentation of a 42-specific γ secretase. *Hum Mol Genet* 2004;**13**:159−70.

24. Glenner GG, Wong CW. Alzheimer's disease and Down's syndrome: sharing of a unique cerebrovascular amyloid fibril protein. *Biochem Biophys Res Commun* 1984;**122**(3):1131−5.

25. Masters CL, Simms G, Weinman NA, Multhaup G, McDonald BL, Beyreuther K. Amyloid plaque core protein in Alzheimer disease and Down syndrome. *PNAS* 1985;**82**:4245−9.

26. Rovelet-Lecrux A, Hannequin D, Raux G, Le Meur N, Laquerrière A, Vital A, et al. APP locus duplication causes autosomal dominant early-onset Alzheimer disease with cerebral amyloid angiopathy. *Nat Genet* 2006;**38**:24−6.

27. Cabrejo L, Guyant-Maréchal L, Laquerrière A, Vercelletto M, De la Fournière F, Thomas-Antérion C, et al. Phenotype associated with APP duplication in five families. *Brain* 2006;**129**:2966−76.

28. Gyure KA, Durham R, Stewart WF, Smialek JE, Troncoso JC. Intraneuronal abeta-amyloid precedes development of amyloid plaques in Down syndrome. *Arch Pathol Lab Med* 2001;**125**:489−92.

29. Citron M, Oltersdorf T, Haass C, McConlogue L, Hung AY, Seubert P, et al. Mutation of the β-amyloid precursor protein in familial Alzheimer's disease increases β-protein production. *Nature* 1992;**360**:672−4.

30. Tokuda T, Fukushima T, Ikeda S, Sekijima Y, Shoji S, Yanagisawa N, et al. Plasma levels of amyloid β proteins Aβ1−40 and Aβ1−42(43) are elevated in Down's syndrome. *Ann Neurol* 1997;**41**:271−3.

31. Mehta PD, Daltonb AJ, Mehtaa SP, Kimc KS, Sersend EA, Wisniewski HM. Increased plasma amyloid β protein 1−42 levels in Down syndrome. *Neurosci Lett* 1998;**241**:13−16.

32. Schupf N, Patela B, Silverman W, Zigman WB, Zhong N, Tycko B, et al. Elevated plasma amyloid β-peptide 1-42 and onset of dementia in adults with Down syndrome. *Neurosci Lett* 2001;**301**:199−203.

33. Iwatsubo T, Mann DMA, Odaka A, Suzuki N, Ihara Y. Amyloid β protein (Aβ) deposition: Aβ42(43) precedes Aβ40 in Down syndrome. *Ann Neurol* 1995;**37**:294−9.

34. Teller JK, Russo C, DeBusk LM, Angelini G, Zaccheo D, Dagna-Bricarelli F, et al. Presence of soluble amyloid β-peptide precedes amyloid plaque formation in Down's syndrome. *Nat Med* 1996;**2**(1):93−5.

35. Lai F. Review. Clinicopathologic features of Alzheimer disease in Down syndrome. *Prog Clin Bio Res* 1992;**379**:15−34.

36. Echeverria V, Cuello AC. Review. Intracellular Aβ amyloid, a sign for worse things to come? *Mol Neurobiol* 2002;**26**(2−3):299−316.

37. Iulita MF, Allard S, Richter L, Munter L-M, Ducatenzeiler A, Weise C, et al. Intracellular Aβ pathology and early cognitive impairments in a transgenic rat overexpressing human amyloid precursor protein: a multidimensional study. *Acta Neuropathol Commun* 2014;**2**:61.

38. Breyhan H, Wirths O, Duan K, Marcello A, Rettig J, Bayer TA. APP/PS1KI bigenic mice develop early synaptic deficits and hippocampus atrophy. *Acta Neuropathol* 2009;**117**:677−85.

39. LaFerla FM, Green KN, Oddo S. Review. Intracellular amyloid-β in Alzheimer's disease. *Nat Rev Neurosci* 2007;**8**:499−509.

40. Bryan KJ, Lee H-G, Perry G, Smith MA. Review. Transgenic mouse models of Alzheimer's disease: behavioral testing and considerations. In: Buccafusco JJ, editor. *Methods of behavior analysis in neuroscience*. 2nd ed. Boca Raton (FL): CRC Press; 2009.
41. Parent A, Linden DJ, Sisodia SS, Borchelt DR. Synaptic transmission and hippocampal long-term potentiation in transgenic mice expressing FAD-linked presenilin 1. *Neurobiol Dis* 1999;**6**:56–62.
42. Schneider I, Reverse D, Dewachter I, Ris L, Caluwaerts N, Kuiperi C, et al. Mutant presenilins disturb neuronal calcium homeostasis in the brain of transgenic mice, decreasing the threshold for excitotoxicity and facilitating long-term potentiation. *J Biol Chem* 2001;**276**:11539–44.
43. Fitzjohn SM, Morton RA, Kuenzi F, Rosahl TW, Shearman M, Lewis H, et al. Age-related impairment of synaptic transmission but normal long-term potentiation in transgenic mice that overexpress the human APP695SWE mutant form of amyloid precursor protein. *J Neurosci* 2001;**21**:4691–8.
44. Hsia AY, Masliah E, McConlogue L, Yu GQ, Tatsuno G, Hu K, et al. Plaque-independent disruption of neural circuits in Alzheimer's disease mouse models. *PNAS* 1999;**96**:3228–33.
45. Moechars D, Dewachter I, Lorent K, Reverse D, Baekelandt V, Naidu A, et al. Early phenotypic changes in transgenic mice that overexpress different mutants of amyloid precursor protein in brain. *J Biol Chem* 1999;**274**:6483–92.
46. Oddo S, Caccamo A, Shepherd JD, Murphy MP, Golde TE, Kayed R, et al. Triple-transgenic model of Alzheimer's disease with plaques and tangles: intracellular Aβ and synaptic dysfunction. *Neuron* 2003;**39**:409–21.
47. Giacchino J, Criado JR, Games D, Henriksen S. In vivo synaptic transmission in young and aged amyloid precursor protein transgenic mice. *Brain Res* 2000;**876**:185–90.
48. Larson J, Lynch G, Games D, Seubert P. Alterations in synaptic transmission and long-term potentiation in hippocampal slices from young and aged PDAPP mice. *Brain Res* 1999;**840**:23–35.
49. Chapman PF, White GL, Jones MW, Cooper-Blacketer D, Marshall VJ, et al. Impaired synaptic plasticity and learning in aged amyloid precursor protein transgenic mice. *Nat Neurosci* 1999;**2**:271–6.
50. Leon WC, Canneva F, Partridge V, Allard S, Ferretti MT, DeWilde A, et al. A novel transgenic rat model with a full Alzheimer's-like amyloid pathology displays pre-plaque intracellular amyloid-β-associated cognitive impairment. *J Alzheimer's Dis* 2010;**20**:113–16.
51. Cattabeni F, Colciaghi F, Di Luca M. Platelets provide human tissue to unravel pathogenic mechanisms of Alzheimer disease. *Prog NeuroPsychopharmacol Biol Psychiatry* 2004;**28**:763–70.
52. <www.alz.org>. [accessed 26.09.15].
53. Corder EH, Saunders AM, Strittmatter WJ, et al. Gene dose of apolipoprotein E type 4 allele and the risk of Alzheimer's disease in late onset families. *Science* 1993;**261**:921–3.
54. Strittmatter WJ. Review. Apolipoprotein E and Alzheimer's disease. *Ann NY Acad Sci* 2000;**924**:91–2.
55. McDowell I. Review. Alzheimer's disease: insights from epidemiology. *Aging* 2001;**13**:143–62.
56. Poirier J. Review. Apolipoprotein E in animal models of CNS injury and in Alzheimer's disease. *Trends Neurosci* 1994;**17**:525–30.
57. Small GW, Ercoli LM, Silverman DH, Huang SC, Komo S, et al. Cerebral metabolic and cognitive decline in persons at genetic risk for Alzheimer's disease. *PNAS* 2000;**97**:6037–42.

58. Mahley RW, Weisgraber KH, Huang Y. Apolipoprotein E4: a causative factor and therapeutic target in neuropathology, including Alzheimer's disease. *PNAS* 2006;**103**:5644—51.

59. Bales KR, Verina T, Dodel RC, et al. Lack of apolipoprotein E dramatically reduces amyloid β-peptide deposition. *Nat Genet* 1997;**17**:263—4.

60. Kivipelto M, Helkala EL, Lasskso MP, et al. Midlife vascular risk factors and Alzheimer's disease in later life: longitudinal, population based study. *BMJ* 2001;**322** (7300):1447—51.

61. Hyman BT, West HL, Rebeck GW, Buldyrev SV, Mantegna RN, Ukleja M, et al. Quantitative analysis of senile plaques in Alzheimer disease: observation of log-normal size distribution and molecular epidemiology of differences associated with apolipoprotein E genotype and trisomy 21 (Down syndrome). *PNAS* 1995;**92**:3586—90.

62. Christensen DZ, Schneuder-Axmann T, Lucassen PJ, Bayer TA, Wirths O. Accumulation of intraneuronal Aβ correlated with ApoE4 genotype. *Acta Neuropathol* 2010;**119**:555—66.

63. Saunders AM, Strittmatter WJ, Schmechel D, St George-Hyslop PH, Pericak-Vance MA, Joo SH, et al. Association of apolipoprotein E allele epsilon 4 with late-onset familial and sporadic Alzheimer's disease. *Neurology* 1993;**43**:1467—72.

64. Mayeux R, Stern Y, Ottman R, Tatemichi TK, Tang MX, Maestre G, et al. The apolipo-protein epsilon 4 allele in patients with Alzheimer's disease. *Ann Neurol* 1993;**34**:752—4.

65. Strittmatter W, Weisgraber K, Huang D, Dong L, Salvesen G, Pericak-Vance M, et al. Binding of human apolipoprotein E to synthetic amyloid β peptide: isoform-specific effects and implications for late-onset Alzheimer disease. *PNAS* 1993;**90**:8098—102.

66. McNamara MJ, Gomez-Isla T, Hyman BT. Apolipoprotein E genotype and deposits of Aβ40 and Aβ42 in Alzheimer disease. *Arch Neurol* 1998;**55**:1001—4.

67. Schmechel DE, Saunders AM, Strittmatter WJ, Crain BJ, Hulette CM, Joo SH, et al. Increased amyloid β-peptide deposition in cerebral cortex as a consequence of apoli-poprotein E genotype in late-onset Alzheimer disease. *PNAS* 1993;**90**:9649—53.

68. Ma J, Yee A, Brewer Jr HB, Das S, Potter H. Amyloid-associated proteins α 1-antichymotrypsin and apolipoprotein E promote assembly of Alzheimer β-protein into filaments. *Nature* 1994;**372**:92—4.

69. Ishii K, Tamaoka A, Mizusawa H, Shoji S, Ohtake T, Fraser PE, et al. Aβ1-40 but not Aβ1-42 levels in cortex correlate with apolipoprotein E epsilon4 allele dosage in spo-radic Alzheimer's disease. *Brain Res* 1997;**748**:250—2.

70. Premkumar DRD, Cohen DL, Hedera P, Friedland RP, Kalaria RN. Apolipoprotein E-epsilon 4 alleles in cerebral amyloid angiopathy and cerebrovascular pathology associated with Alzheimer's disease. *Am J Pathol* 1996;**148**:2083—95.

71. Kok E, Haikonen S, Luoto T, et al. Apolipoprotein E-dependent accumulation of Alzheimer disease-related lesions begins in middle age. *Ann Neurol* 2009;**65**:650—7.

72. Zhao L, Lin S, Bales KR, et al. Macrophage-mediated degradation of β-amyloid via an apolipoprotein E isoform-dependent mechanism. *J Neurosci* 2009;**29**:3603—12.

73. Pillot T, Goethals M, Najib J, Labeur C, Lins L, Chambaz J, et al. Beta-amyloid peptide interacts specifically with the carboxy-terminal domain of human apolipoprotein E: relevance to Alzheimer's disease. *J Neurochem* 1999;**72**:230—7.

74. Sanan DA, Weisgraber KH, Russell SJ, Mahley RW, Huang D, Saunders A, et al. Apolipoprotein E associates with β amyloid peptide of Alzheimer's disease to form novel monofibrils. *J Clin Invest* 1994;**94**:860—9.

75. Wisniewski T, Castaño EM, Golabek A, Vogel T, Frangione B. Acceleration of Alzheimer's fibril formation by apolipoprotein E in vitro. *Am J Pathol* 1994;**145**:1030—5.

76. Royston M, Mann D, Pickering-Brown S, Owen F, Perry R, Raghaven R, et al. Apolipoprotein E ε2 allele promotes longevity and protects patients with Down's syndrome from dementia. *NeuroReport* 1994;**5**:2583−5.
77. Schupf N, Kapell D, Lee JH, Zigman W, Canto B, Tycko B, et al. Onset of dementia is associated with apolipoprotein E epsilon 4 in Down's syndrome. *Ann Neurol* 1996;**40**:799−801.
78. Flood JF, Morley JE, Roberts E. Amnestic effects in mice of four synthetic peptides homologous to amyloid-β protein from patients with Alzheimer disease. *PNAS* 1991;**88**:3363−6.
79. Cleary JP, Walsh DM, Hofmeister JJ, Shankar GM, Kuskowski MA, Ashe KH. Natural oligomers of the amyloid-β protein specifically disrupt cognitive function. *Nat Neurosci* 2005;**8**:79−84.
80. Maurice T, Lockhart BP, Privat A. Amnesia induced in mice by centrally administered β-amyloid peptides involves cholinergic dysfunction. *Brain Res* 1996;**706**:181−93.
81. Sipos E, Kurunczi A, Kasza A, Horvath J, Felszeghy K, Laroche S, et al. Beta-Amyloid pathology in the entorhinal cortex of rats induces memory deficits: implications for Alzheimer's disease. *Neuroscience* 2007;**147**:28−36.
82. Walsh DM. Review. Aβ oligomers-a decade of discovery. *J Neurochem* 2007;**101**:1172−84.
83. Irvine GB, El-Agnaf OM, Shankar GM, Walsh DM. Review. Protein aggregation in the brain: the molecular basis for Alzheimer's and Parkinson's diseases. *Mol Med* 2008;**14**:451−64.
84. Shankar GM, Bloodgood BL, Townsend M, Walsh DM, Sabatini BL. Natural oligomers of the Alzheimer amyloid-β protein induce reversible synapse loss by modulating an NMDA-type glutamate receptor-dependent signaling pathway. *J Neurosci* 2007;**27**:2866−75.
85. Hyman BT, Gomez-Isla T. Review. The natural history of Alzheimer neurofibrillary tangles and amyloid deposits. *Neurobiol Aging* 1997;**18**:386−7.
86. Karran E, Mercken M, De Strooper B. Review. The amyloid cascade hypothesis for Alzheimer's disease: an appraisal for the development of therapeutics. *Nat Rev Drug Discov* 2011;**10**:698−712.
87. Evin G, Li Q-X. Review. Platelets and Alzheimer's disease: potential of APP as a biomarker. *World J Psychiatr* 2012;**2**(6):102−13.
88. Hardy J, Selkoe DJ. The amyloid hypothesis of Alzheimer's disease: progress and problems on the road to therapeutics. *Science* 2002;**297**:353−6.
89. Ferreira ST, Klein WL. Review. The Aβ oligomer hypothesis for synapse failure and memory loss in Alzheimer's disease. *Neurobiol Learn Mem* 2011;**96**:529−43.
90. Castellani RJ, Smith MA. Review. Compounding artefacts with uncertainty, and an amyloid cascade hypothesis that is 'too big to fail'. *J Pathol* 2011;**224**(2):147−52.
91. Koudinov AR, Berezov TT. Review. Alzheimer's amyloid-β (Aβ) is an essential synaptic protein, not neurotoxic junk. *Acta Neurobiol Exp* 2004;**64**:71−9.
92. D'Andrea MR, Nagele RG, Wang H-Y, Peterson PA, Lee DHS. Evidence that neurones accumulating amyloid can undergo lysis to form amyloid plaques in AD. *Histopathology* 2001;**38**:120−34.
93. D'Andrea MR, Cole GM, Ard MD. The microglial phagocytic role with specific plaque types in the Alzheimer disease brain. *Neurobiol Aging* 2004;**25**:675−83.
94. D'Andrea MR, Reiser PA, Polkovitch DA, Branchide B, Hertzog BH, Belkowski S, et al. The use of formic acid to embellish amyloid plaque detection in Alzheimer's disease tissues misguides key observations. *Neurosci Lett* 2003;**342**:114−18.
95. Lansdall CL. Review. An effective treatment of Alzheimer's disease must consider both amyloid and tau. *Biosci Horiz* 2014;**7**:1.

96. Schmitz C, Rutten BPF, Pielen A, Schafer S, Wirths O, Tremp G, et al. Hippocampal neuron loss exceeds amyloid plaque load in a transgenic mouse model of Alzheimer's disease. *Am J Pathol* 2004;**164**:1495–502.
97. D'Andrea MR. *Bursting neurons and fading memories*. New York: Elsevier; 2014.
98. D'Andrea MR, Nagele RG, Wang H-Y, Peterson PA, Lee DHS. Origin of an amyloid plaque in AD. *Soc Neurosci Abstr* 1999;25.
99. D'Andrea MR, Reiser PA, Gumula NA, Hertzog BM, Andrade-Gordon R. Application of triple immunohistochemistry to characterize amyloid plaque-associated inflammation in brains with Alzheimer's disease. *Biotech Histochem* 2001;**76**(2):97–106.
100. Parri R. Book review: bursting neurons and fading memories. An alternative hypothesis of the pathogenesis of Alzheimer's disease. *Biotech Histochem* 2015;**90**(6):495–6.
101. Wirths O, Multhaup G, Bayer TA. Review. A modified β-amyloid hypothesis: intraneuronal accumulation of the β-amyloid peptide — the first step of a fatal cascade. *J Neurochem* 2004;**91**:513–20.
102. Knobloch M, Konietzko U, Krebs DC, Nitsch RM. Intracellular Aβ and cognitive deficits precede-amyloid deposition in transgenic arcAβ mice. *Neurobiol Aging* 2007;**28**:1297–306.
103. D'Andrea MR, Lee DHS, Wang H-Y, Nagele RG. Targeting intracellular Aβ42 for Alzheimer's disease drug discovery. *Drug Dev Res* 2002;**56**:194–200.
104. Gouras GK, Tsai J, Näslund J, Vincent B, Edgar M, Checler F, et al. Intraneuronal Aβ42 accumulation in human brain. *Am J Path* 2000;**156**(1):15–20.
105. D'Andrea MR, Nagele RG. Morphologically distinct types of amyloid plaques point the way to a better understanding of Alzheimer's disease pathogenesis. *Biotech Histochem* 2010;**85**(2):133–47.
106. Marchesi VT. Review. An alternative interpretation of the amyloid Aβ hypothesis with regard to the pathogenesis of Alzheimer's disease. *PNAS* 2005;**102**(26):9093–8.
107. Chui DH, Tanahashi H, Ozawa K, Ikeda S, Checler F, Ueda O, et al. Transgenic mice with Alzheimer presenilin 1 mutations show accelerated neurodegeneration without amyloid plaque formation. *Nat Med* 1999;**5**(5):560–4.
108. Mohamed A, de Chaves EP. Review. Aβ internalization by neurons and glia. *Int J Alzheimer's Dis* 2011;**127984**:17.
109. Wirths O, Bayer TA. Review. Intraneuronal Aβ accumulation and neurodegeneration: lessons from transgenic models. *Life Sci* 2012;**91**:1148–52.
110. Bayer TA, Schafer S, Breyhan H, Wirths O, Treiber C, Multhaup G. Review. A vicious circle: role of oxidative stress, intraneuronal Aβ and Cu in Alzheimer's disease. *Clin Neuropathol* 2006;**25**(6):163–71.
111. Billings LM, Oddo S, Green KN, McGaugh JL, LaFerla FM. Intraneuronal Aβ causes the onset of early Alzheimer's disease-related cognitive deficits in transgenic mice. *Neuron* 2005;**45**:675–88.
112. D'Andrea MR, Nagele RG. MAP-2 immunolabeling can distinguish diffuse from dense-core amyloid plaques in brains with Alzheimer's disease. *Biotech Histochem* 2002;**77**(2):95–103.
113. Takahashi RH, Capetillo-Zarate E, Lin MT, Milner TA, Gouras GK. Accumulation of intraneuronal β-amyloid 42 peptides is associated with early changes in microtubule-associated protein 2 in neurites and synapses. *PLoS One* 2013;**8**(1):e51965.
114. Capetillo-Zarate E, Staufenbiel M, Abramowski D, Haass C, Escher A, et al. Selective vulnerability of different types of commissural neurons for amyloid β-protein-induced neurodegeneration in APP23 mice correlates with dendritic tree morphology. *Brain* 2006;**129**:2992–3005.
115. Moolman DL, Vitolo OV, Vonsattel JP, Shelanski ML. Dendrite and dendritic spine alterations in Alzheimer models. *J Neurocytol* 2004;**33**:377–87.

116. Wu C-C, Chawla F, Games D, Rydel RE, Freedman S, et al. Selective vulnerability of dentate granule cells prior to amyloid deposition in PDAPP mice: digital morphometric analyses. *PNAS* 2004;**101**:7141−6.

117. Canas PM, Porciuncula LO, Cunha GMA, Silva CG, Machado NJ, et al. Adenosine A2A receptor blockade prevents synaptotoxicity and memory dysfunction caused by β-amyloid peptides via p38 mitogen-activated protein kinase pathway. *J Neurosci* 2009;**29**:14741−51.

118. Dziewczapolski G, Glogowski CM, Masliah E, Heinemann SF. Deletion of the α7 nicotinic acetylcholine receptor gene improves cognitive deficits and synaptic pathology in a mouse model of Alzheimer's disease. *J Neurosci* 2009;**29**:8805−15.

119. Echeverria V, Ducatenzeiler A, Dowd E, Jänne J, Grant SM, Szyf M, et al. Altered mitogen-activated protein kinase signaling, tau hyperphosphorylation and mild spatial learning dysfunction in transgenic rats expressing the β-amyloid peptide intracellularly in hippocampal and cortical neurons. *Neuroscience* 2004;**129**(3):583−92.

120. Takahashi RH, Capetillo-Zarate E, Lin MT, Milner TA, Gouras GK. Co-occurrence of Alzheimer's disease β-amyloid and tau pathologies at synapses. *Neurobiol Aging* 2010;**31**(7):1145−52.

121. Parihar MS, Brewer GJ. Review. Amyloid beta as a modulator of synaptic plasticity. *J Alzheimer's Dis* 2010;**22**(3):741−63.

122. Rank KB, Pauley AM, Bhattacharya K, Wang Z, Evans DB, Fleck TJ, et al. Direct interaction of soluble human recombinant tau protein with Aβ1−42 results in tau aggregation and hyperphosphorylation by tau protein kinase II. *FEBS Lett* 2002;**514**:263−8.

123. Stancu I-C, Vasconcelos B, Terwel D, Dewachter I. Review. Models of β-amyloid induced Tau-pathology: the long and "folded" road to understand the mechanism. *Mol Neurodegener* 2014;**9**:51.

124. Shipton OA, Leitz JR, Dworzak J, Acton CE, Tunbridge EM, Denk F, et al. Tau protein is required for amyloid (beta)-induced impairment of hippocampal long-term potentiation. *J Neurosci* 2011;**31**(5):1688−92.

125. Munoz-Montano JR, Moreno FJ, Avila J, Diaz-Nido J. Lithium inhibits Alzheimer's disease-like tau protein phosphorylation in neurons. *FEBS Lett* 1997;**411**:183−8.

126. Busciglio J, Lorenzo A, Yeh J, Yankner BA. Beta-amyloids fibrils induce tau phosphorylation and loss of microtubule binding. *Neuron* 1995;**14**:879−88.

127. Terwel D, Muyllaert D, Dewachter I, Borghgraef P, Croes S, Devijver H, et al. Amyloid activates GSK-3β to aggravate neuronal tauopathy in bigenic mice. *Am J Pathol* 2008;**172**(3):786−98.

128. Hurtado DE, Molina-Porcel L, Carroll JC, Macdonald C, Aboagye AK, Lee VM. Selectively silencing GSK-3 isoforms reduces plaques and tangles in mouse models of Alzheimer's disease. *J Neurosci* 2012;**32**(21):7392−402.

129. Lewis J, Dickson DW, Lin WL, Chisholm L, Corral A, Jones G, et al. Enhanced neurofibrillary degeneration in transgenic mice expressing mutant tau and APP. *Science* 2001;**293**:1487−91.

130. Perez M, Cuadros R, Benitez MJ, Jimnez JS. Interaction of Alzheimer's disease amyloid β peptide fragment 25−35 with tau protein, and with a tau peptide containing the microtubule binding domain. *J Alzheimer's Dis* 2004;**6**:461−7.

131. Oliveira DL, Rizzo J, Joffe LS, Godinho RMC, Rodrigues ML. Where do they come from and where do they go: candidates for regulating extracellular vesicle formation in fungi. *Int J Mol Sci* 2013;**14**:9581−603.

132. Zerbinatti CV, Wahrle SE, Kim H, Cam JA, Bales K, Paul SM, et al. Apolipoprotein E and low density lipoprotein receptor-related protein facilitate intraneuronal Aβ42 accumulation in amyloid model mice. *J Bio Chem* 2006;**281**(47):36180−6.

133. Cataldo AM, Petanceska S, Terio NB, Peterhoff CM, Durham R, Mercken M, et al. Aβ localization in abnormal endosomes: association with earliest Aβ elevations in AD and Down syndrome. *Neurobiol Aging* 2004;**25**:1263–72.

134. Langui D, Girardot N, El Hachimi KH, Allinquant B, Blanchard V, Pradier L, et al. Subcellular topography of neuronal Aβ peptide in APPxPS1 transgenic mice. *Am J Pathol* 2004;**165**:1465–77.

135. Takahashi RH, Milner TA, Li F, Nam EE, Edgar MA, Yamaguchi H, et al. Intraneuronal Alzheimer Aβ42 accumulates in multivesicular bodies and is associated with synaptic pathology. *Am J Pathol* 2002;**161**:1869–79.

136. Shie FS, LeBoeur RC, Jin LW. Early intraneuronal Aβ deposition in the hippocampus of APP transgenic mice. *NeuroReport* 2003;**14**:123–9.

137. Zheng L, Cedazo-Minguez A, Hallbeck M, Jerhammar F, Marcusson J, Terman A. Intracellular distribution of amyloid beta peptide and its relationship to the lysosomal system. *Transl Neurodegener* 2012;**1**:19.

138. Zheng L, Roberg K, Jerhammar F, Marcusson J, Terman A. Autophagy of amyloid β-protein in differentiated neuroblastoma cells exposed to oxidative stress. *Neurosci Lett* 2006;**394**(3):184–9.

139. Vekrellis K, Ye Z, Qiu WQ, Walsh D, Hartley D, Chesneau V, et al. Neurons regulate extracellular levels of amyloid β-protein via proteolysis by insulin-degrading enzyme. *J Neurosci* 2000;**20**:1657–65.

140. Iwata N, Tsubuki S, Takaki Y, Shirotani K, Lu B, Gerard NP, et al. Metabolic regulation of brain Aβ by neprilysin. *Science* 2001;**292**:1550–2.

141. Stine Jr WB, Dahlgren KN, Krafft GA, LaDu MJ. Subcellular topography of neuronal Aβ peptide in APPxPS1 transgenic mice. *J Biol Chem* 2003;**278**:11612–22.

142. Pasternak SH, Callahan JW, Mahuran DJ. Review. The role of the endosomal/lysosomal system in amyloid-β production and the pathophysiology of Alzheimer's disease: reexamining the spatial paradox from a lysosomal perspective. *J Alzheimer's Dis* 2004;**6**:53–65.

143. Knauer MF, Soreghan B, Burdick D, Kosmoski J, Glabe CG. Intracellular accumulation and resistance to degradation of the Alzheimer amyloid A4/β protein. *PNAS* 1992;**89**:7437–41.

144. Bahr BA, Hoffman KB, Yang AJ, Hess US, Glabe CG, Lynch G. Amyloid β protein is internalized selectively by hippocampal field CA1 and causes neurons to accumulate amyloidogenic carboxy terminal fragments of the amyloid precursor protein. *J Comp Neurol* 1998;**397**:139–47.

145. Levi O, Dolev I, Belinson H, Michaelson DM. Intraneuronal amyloid-β plays a role in mediating the synergistic pathological effects of apoE4 and environmental stimulation. *J Neurochem* 2007;**103**:1031–40.

146. Belinson H, Lev D, Masliah E, Michaelson DM. Activation of the amyloid cascade in apolipoprotein E4 transgenic mice induces lysosomal activation and neurodegeneration resulting in marked cognitive deficits. *J Neurosci* 2008;**28**:4690–701.

147. Boland B, Kumar A, Lee S, Platt FM, Wegiel J, Yu WH, et al. Autophagy induction and autophagosome clearance in neurons: relationship to autophagic pathology in Alzheimer's disease. *J Neurosci* 2008;**28**:6926–37.

148. Tate BA, Mathews PM. Targeting the role of the endosome in the pathophysiology of Alzheimer's disease: a strategy for treatment. *Sci Aging Knowledge Environ* 2006;**10**:re2.

149. Nagele RG, D'Andrea MR, Anderson WJ, Wang H-Y. Intracellular accumulation of β-amyloid$_{1-42}$ in neurons is facilitated by the α7 nicotinic acetylcholine receptor in Alzheimer's disease. *Neuroscience* 2002;**110**(2):199–211.

150. Bahr BA, Bendiske J. Review. The neuropathogenic contributions of lysosomal dysfunction. *J Neurochem* 2002;**83**:481–9.

151. Callahan LM, Vaules WA, Coleman PD. Quantitative decrease in synaptophysin message expression and increase in cathepsin D message expression in Alzheimer disease neurons containing neurofibrillary tangles. *J Neuropathol Exp Neurol* 1999;**58**:275−87.

152. Bendiske J, Bahr BA. Lysosomal activation is a compensatory response against protein accumulation and associated synaptopathogenesis—an approach for slowing Alzheimer disease? *J Neuropathol Exp Neurol* 2003;**62**:451−63.

153. Bendiske J, Caba E, Brown QB, Bahr BA. Intracellular deposition, microtubule destabilization, and transport failure: an "early" pathogenic cascade leading to synaptic decline. *J Neuropathol Exp Neurol* 2002;**61**:640−50.

154. Kanju PM, Parameshwaran K, Vaithianathan T, Sims CM, Huggins K, Bendiske J, et al. Lysosomal dysfunction produces distinct alterations in synaptic α-amino-3-hydroxy-5-methylisoxazolepropionic acid and N-methyl-D-aspartate receptor currents in hippocampus. *J Neuropathol Exp Neurol* 2007;**66**:779−88.

155. Cataldo AM, Barnett JL, Pieroni C, Nixon RA. Increased neuronal endocytosis and protease delivery to early endosomes in sporadic Alzheimer's disease: neuropathologic evidence for a mechanism of increased β-amyloidogenesis. *J Neurosci* 1997;**17**:6142−51.

156. Ohyagi Y, Asahara H, Chui D-H, Tsuruta Y, Sakae N, Miyoshi K, et al. Intracellular Aβ42 activates p53 promoter: a pathway to neurodegeneration in Alzheimer's disease. *FASEB J* 2005;**19**(2):255−7.

157. Bückig A, Tikkanen R, Herzog V, Schmits A. Cytosolic and nuclear aggregation of the amyloid β-peptide following its expression in the endoplasmic reticulum. *Histochem Cell Biol* 2002;**118**:353−60.

158. Lopez Salon M, Pasquini L, Besio Moreno M, Pasquini JM, Soto E. Relationship between β-amyloid degradation and the 26S proteasome in neural cells. *Exp Neurol* 2003;**180**:131−43.

159. Lopez Salon M, Morelli L, Castaño EM, Soto EF, Pasquini JM. Defective ubiquitination of cerebral proteins in Alzheimer's disease. *J Neurosci Res* 2000;**62**:302−10.

160. De Vrij FMS, Sluijs JA, Gregori L, Fischer DF, Hermens WT, Goldgaber D, et al. Mutant ubiquitin expressed in Alzheimer's disease causes neuronal death. *FASEB J* 2001;**15**:2680−8.

161. Oh S, Hong HS, Hwang E, Sim HJ, Lee W, Shin SJ, et al. Amyloid peptide attenuates the proteasome activity in neuronal cells. *Mech Ageing Dev* 2005;**126**(12):1292−9.

162. Gruenberg J. Review. The endocytic pathway: a mosaic of domains. *Nat Rev Mol Cell Biol* 2001;**2**:721−30.

163. Almeida CG, Takahashi RH, Gouras GK. Beta-amyloid accumulation impairs multivesicular body sorting by inhibiting the ubiquitin-proteasome system. *J Neurosci* 2006;**26**(16):4277−88.

164. Takahashi RH, Almeida CG, Kearney PF, Yu F, Lin MT, Milner TA, et al. Oligomerization of Alzheimer's β-amyloid within processes and synapses of cultured neurons and brain. *J Neurosci* 2004;**24**:3592−9.

165. Fay DS, Fluet A, Johnson CJ, Link CD. In vivo aggregation of β-amyloid peptide variants. *J Neurochem* 1998;**71**:1616−25.

166. Link CD, Johnson CJ, Fonte V, Paupard M, Hall DH, Styren S, et al. Visualization of fibrillar amyloid deposits in living, transgenic *Caenorhabditis elegans* animals using the sensitive amyloid dye, X-34. *Neurobiol Aging* 2001;**22**:217−26.

167. Fonte V, Kapulkin V, Taft A, Fluet A, Friedman D, Link CD. Interaction of intracellular β amyloid peptide with chaperone proteins. *PNAS* 2002;**99**(14):9439−44.

168. Ma J, Lindquist S. Wild-type PrP and a mutant associated with prion disease are subject to retrograde transport and proteasome degradation. *PNAS* 2001;**98**:14955−60.

169. Hamos JE, DeGennaro LJ, Drachman DA. Synaptic loss in Alzheimer's disease and other dementias. *Neurology* 1989;**39**:355−61.

170. Masliah E, Terry RD, DeTeresa RM, Hansen LA. Immunohistochemical quantification of the synapse-related protein synaptophysin in Alzheimer disease. *Neurosci Lett* 1989;**103**:234–9.
171. Mesulam MM. Neuroplasticity failure in Alzheimer's disease: bridging the gap between plaques and tangles. *Neuron* 1999;**24**:521–9.
172. Jacobsen JS, Wu CC, Redwine JM, Comery TA, Arias R, et al. Early-onset behavioral and synaptic deficits in a mouse model of Alzheimer's disease. *PNAS* 2006;**103**:5161–6.
173. Terry RD, Masliah E, Salmon DP, Butters N, DeTeresa R, Hill R, et al. Physical basis of cognitive alterations in Alzheimer's disease: synapse loss is the major correlate of cognitive impairment. *Ann Neurol* 1991;**30**:572–80.
174. Masliah E, Mallory M, Alford M, DeTeresa R, Hansen LA, McKeel Jr. DW, et al. Altered expression of synaptic proteins occurs early during progression of Alzheimer's disease. *Neurology* 2001;**56**:127–9.
175. Yuste R, Bonhoeffer T. Review. Genesis of dendritic spines: insights from ultrastructural and imaging studies. *Nat Rev Neurosci* 2004;**5**:24–34.
176. Harris KM, Kater SB. Review. Dendritic spines: cellular specializations imparting both stability and flexibility to synaptic function. *Annu Rev Neurosci* 1994;**17**:341–71.
177. Carlisle HJ, Kennedy MB. Review. Spine architecture and synaptic plasticity. *Trends Neurosci* 2005;**28**:182–7.
178. Segal M. Review. Dendritic spines and long-term plasticity. *Nat Rev Neurosci* 2005;**6**:277–84.
179. Ferrer I, Gullotta F. Down's syndrome and Alzheimer's disease: dendritic spine counts in the hippocampus. *Acta Neuropathol* 1990;**79**:680–5.
180. Spires TL, Meyer-Luehmann M, Stern EA, McLean PJ, Skoch J, Nguyen PT, et al. Dendritic spine abnormalities in amyloid precursor protein transgenic mice demonstrated by gene transfer and intravital multiphoton microscopy. *J Neurosci* 2005;**25**:7278–87.
181. Sze C, Bi H, Kleinschmidt-DeMasters BK, Filley CM, Martin LJ. *N*-Methyl-D-aspartate receptor subunit proteins and their phosphorylation status are altered selectively in Alzheimer's disease. *J Neurol Sci* 2001;**182**:151–9.
182. Mishizen-Eberz AJ, Rissman RA, Carter TL, Ikonomovic MD, Wolfe BB, Armstrong DM. Biochemical and molecular studies of NMDA receptor subunits NR1/2A/2B in hippocampal subregions throughout progression of Alzheimer's disease pathology. *Neurobiol Dis* 2004;**15**:80–92.
183. Harigaya Y, Shoji M, Shirao T, Hirai S. Disappearance of actin-binding protein, drebrin, from hippocampal synapses in Alzheimer's disease. *J Neurosci Res* 1996;**43**:87–92.
184. Wang H, Megill A, He K, Kirkwood A, Lee H-K. Consequences of inhibiting amyloid precursor protein processing enzymes on synaptic function and plasticity. *Neural Plast* 2012;**272374**:24.
185. Tu S, Okamoto S-I, Lipton SA, Xu H. Review. Oligomeric Aβ-induced synaptic dysfunction in Alzheimer's disease. *Mol Neurodegener* 2014;**9**:48.
186. Tampellini D, Rahman N, Gallo EF, Huang Z, Dumont M, Capetillo-Zarate E, et al. Synaptic activity reduces intraneuronal Aβ, promotes APP transport to synapses and protects against Aβ-related synaptic alterations. *J Neurosci* 2009;**29**(31):9704–13.
187. Hooli BV, Tanzi RE. Review. A current view of Alzheimer's disease. *Bio Rep* 2009;**1**:54.
188. Schmitt TL, Steiner E, Trieb K, Grubeck-Loebenstein B. Amyloid β-protein$_{25-35}$ increases cellular APP and inhibits the secretion of APPs in human extraneuronal cells. *Exp Cell Res* 1997;**234**:336–40.
189. Wei W, Nguyen LN, Kessels HW, Hagiwara H, Sisodia S, Malinow R. Amyloid β from axons and dendrites reduces local spine number and plasticity. *Nat Neurosci* 2010;**13**:190–6.

190. Walsh DM, Tseng BP, Rydel RE, Podlisny MB, Selkoe DJ. The oligomerization of amyloid β-protein begins intracellularly in cells derived from human brain. *Biochemistry* 2000;**39**:10831−9.
191. Jin M, Shepardson N, Yang T, Chen G, Walsh D, Selkoe DJ. Soluble amyloid β-protein dimers isolated from Alzheimer cortex directly induce tau hyperphosphorylation and neuritic degeneration. *PNAS* 2011;**108**:5819−24.
192. Walsh DM, Klyubin I, Fadeeva JV, Cullen WK, Anwyl R, Wolfe MS, et al. Naturally secreted oligomers of amyloid β protein potently inhibit hippocampal long-term potentiation in vivo. *Nature* 2002;**416**:535−9.
193. Shankar GM, Li S, Mehta TH, Garcia-Munoz A, Shepardson NE, Smith I, et al. Review. Amyloid-β protein dimers isolated directly from Alzheimer's brains impair synaptic plasticity and memory. *Nat Med* 2008;**14**:837−42.
194. Tampellini D, Gouras GK. Review. Synapses, synaptic activity and intraneuronal Aβ in Alzheimer's disease. *Front Aging Neurosci* 2010;**2**(13):1.
195. Bayer TA, Wirths O. Review. Intracellular accumulation of amyloid-β — a predictor for synaptic dysfunction and neuron loss in Alzheimer's disease. *Front Aging Neurosci* 2010;**2**(8):10.
196. Li YP, Bushnell AF, Lee CM, Perlmutter LS, Wong SK. Beta-amyloid induces apoptosis in human-derived neurotypic SH-SY5Y cells. *Brain Res* 1996;**738**:196−204.
197. Li F, Calingasan NY, Yu F, et al. Increased plaque burden in brains of APP mutant MnSOD heterozygous knockout mice. *J Neurochemistry* 2004;**89**(5):1308−12.
198. Moreno H, Yu E, Pigino G, Hernandez AI, Kim N, Moreira JE, et al. Synaptic transmission block by presynaptic injection of oligomeric amyloid beta. *PNAS* 2009;**106**:5901−6.
199. Masters CL, Multhaup G, Simms G, Pottgiesser J, Martins RN, Beyreuther K. Neuronal origin of a cerebral amyloid: neurofibriliary tangles of Alzheimer's disease contain the same protein as the amyloid of plaque cores and blood vessels. *EMBO J* 1985;**4**(11):2757−63.
200. Picone P, Nuzzo D, Caruana L, Scafidi V, Di Carlo M. Mitochondrial dysfunction: different routes to Alzheimer's disease therapy. *Oxid Med Cell Longev* 2014;**780179**:11.
201. Eckert A, Keil U, Marques CA, Bonert A, Frey C, Schüssel K, et al. Review. Mitochondrial dysfunction, apoptotic cell death, and Alzheimer's disease. *Biochem Pharmacol* 2003;**66**:1627−34.
202. Moreira PI, Carvalho C, Zhu X, Smith MA, Perry G. Mitochondrial dysfunction is a trigger of Alzheimer's disease pathophysiology. *Biochim Biophys Acta* 2010;**1802** (1):2−10.
203. Swerdlow RH, Burns JM, Khan SM. Review. The Alzheimer's disease mitochondrial cascade hypothesis. *J Alzheimer's Dis* 2010;**20**(2):S265−79.
204. Manfredi JJ. Review. P53 and apoptosis: it's not just in the nucleus anymore. *Mol Cell* 2003;**11**:552−4.
205. Mihara M, Erster S, Zaika A, Petrenko O, Chittenden T, Pancoska P, et al. P53 has a direct apoptogenic role at the mitochondria. *Mol Cell* 2003;**11**:577−90.
206. Dumont P, Leu JI-Ju, Della Pietra III AC, George DL, Murphy M. The codon 72 polymorphic variants of p53 have markedly different apoptotic potential. *Nat Genet* 2003;**33**:357−65.
207. Gilman CP, Chan SL, Guo Z, Zhu X, Greig N, Mattson MP. P53 is present in synapses where it mediates mitochondrial dysfunction and synaptic degeneration in response to DNA damage, and oxidative and excitotoxic insults. *Neuromol Med* 2003;**3**:159−72.
208. Chen X, Petranovic D. Amyloid-β peptide-induced cytotoxicity and mitochondrial dysfunction in yeast. *FEMS Yeast Res* 2015;**15**:6.
209. Hardingham GE, Bading H. Review. The Yin and Yang of NMDA receptor signalling. *Trends Neurosci* 2003;**26**:81−9.

210. Zucker RS, Regehr WG. Review. Short-term synaptic plasticity. *Annu Rev Physiol* 2002;**64**:355–405.

211. Wang HY, Lee DH, Davis CB, Shank RP. Amyloid peptide Aβ(1–42) binds selectively and with picomolar affinity to α7 nicotinic acetylcholine receptors. *J Neurochem* 2000;**75**:1155–61.

212. Dougherty JJ, Wu J, Nichols RA. Beta-amyloid regulation of presynaptic nicotinic receptors in rat hippocampus and neocortex. *J Neurosci* 2003;**23**:6740–7.

213. Puzzo D, Privitera L, Leznik E, Fa M, Staniszewski A, Palmeri A, et al. Picomolar amyloid β positively modulates synaptic plasticity and memory in hippocampus. *J Neurosci* 2008;**28**:14537–45.

214. LaFerla FM. Review. Calcium dyshomeostasis and intracellular signalling in Alzheimer's disease. *Nat Rev Neurosci* 2002;**3**:862–72.

215. Zhang SJ, Zou M, Lu L, Lau D, Ditzel DA, Delucinge-Vivier C, et al. Nuclear calcium signaling controls expression of a large gene pool: identification of a gene program for acquired neuroprotection induced by synaptic activity. *PLoS Genet* 2009;**5**: e1000604.

216. Yamamoto-Sasaki M, Ozawa H, Saito T, Rosler M, Riederer P. Impaired phosphorylation of cyclic AMP response element binding protein in the hippocampus of dementia of the Alzheimer type. *Brain Res* 1999;**824**:300–3.

217. Hardingham GE, Arnold FJ, Bading H. Nuclear calcium signaling controls CREB-mediated gene expression triggered by synaptic activity. *Nat Neurosci* 2001;**4**:261–7.

218. Toescu EC, Verkhratsky A, Landfield PW. Ca2 + regulation and gene expression in normal brain aging. *Trends Neurosci* 2004;**27**:614–20.

219. Davies P, Maloney AJ. Selective loss of central cholinergic neurons in Alzheimer's disease. *Lancet* 1976;**2**:1403.

220. Perry EK, Perry RH, Blessed G, Tomlinson BE. Necropsy evidence of central cholinergic deficits in senile dementia. *Lancet* 1977;**1**:189.

221. Kaeser MD, Iggo RD. Chromatin immunoprecipitation analysis fails to support the latency model for regulation of p53 DNA binding activity in vivo. *PNAS* 2002;**99**:95–100.

222. Lassmann H, Bancher C, Breitschopf H, Wegiel J, Bobinski M, Jellinger K, et al. Cell death in Alzheimer's disease evaluated by DNA fragmentation in situ. *Acta Neuropathol* 1995;**89**:35–41.

223. Smale G, Nichols NR, Brady DR, Finch CE, Horton Jr. WE. Evidence for apoptotic cell death in Alzheimer's disease. *Exp Neurol* 2005;**133**:225–30.

224. Chui DH, Dobo E, Makifuchi T, Akiyama H, Kawakatsu S, Petit A, et al. Apoptotic neurons in Alzheimer's disease frequently show intracellular Aβ42 labeling. *J Alzheimer's Dis* 2001;**3**:231–9.

225. Norbury CJ, Zhivotovski B. Review. DNA damage-induced apoptosis. *Oncogene* 2004;**23**:2797–808.

226. Ohyagi Y, Yamada T, Nishioka K, Clarke NJ, Tomlinson AJ, Naylor S, et al. Selective increase in cellular Aβ42 is related to apoptosis but not to necrosis. *NeuroReport* 2000;**11**:167–71.

227. Guo Q, Christakos S, Robinson N, Mattson MP. Calbindin D28k blocks the proapoptotic actions of mutant presenilin 1: reduced oxidative stress and preserved mitochondrial function. *PNAS* 1998;**95**:3227–32.

228. D'Andrea MR. Evidence linking neuronal cell death to autoimmunity in Alzheimer's disease. *Brain Res* 2003;**982**:19–30.

229. Kienlen-Campard P, Miolet S, Tasiaux B, Octave J-N. Intracellular amyloid-β1-42, but not extracellular soluble amyloid-β peptides, induces neuronal apoptosis. *J Biol Chem* 2002;**277**:15666–70.

230. Chui DH, Tanahashi H, Ozawa K, Ifeda S, Checler F, Ueda O, et al. Transgenic mice with Alzheimer presenilin 1 mutations show accelerated neurodegeneration without amyloid plaque formation. *Nat Med* 1999;**5**:560−4.
231. De la Monte SM, Sohn YK, Wands JP. Correlates of p53- and Fas (CD95)-mediated apoptosis in Alzheimer's disease. *J Neurol Sci* 1997;**152**:73−83.
232. Slack RS, Belliveau DJ, Rosenberg M, Atwal J, Lochmuller H, Aloyz R. Adenovirus-mediated gene transfer of the tumor suppressor, p53, induces apoptosis in postmitotic neurons. *J Cell Biol* 1996;**135**:1085−96.
233. Xiang H, Hochman DW, Saya H, Fujiwara T, Schwartzkroin PA, Morrison RS. Evidence for p53-mediated modulation of neuronal viability. *J Neurosci* 1996;**16**:6753−65.
234. Hughes PE, Alexi T, Dragunow M. A role for the tumour suppressor gene p53 in regulating neuronal apoptosis. *NeuroReport* 1997;**8**:v−xii.
235. Seidl R, Fang-Kircher S, Bidmon B, Cairns N, Lubec G. Apoptosis-associated proteins p53 and APO-1/Fas (CD95) in brains of adult patients with Down syndrome. *Neurosci Lett* 1999;**260**:9−12.
236. Zhang Y, McLaughlin R, Goodyer C, LeBlanc A. Selective cytotoxicity of intracellular amyloid β peptide1-42 through p53 and Bax in cultured primary human neurons. *J Cell Biol* 2002;**156**:519−29.
237. Sherrington R, Rogaev EI, Liang Y, Rogaeva EA, Levesque G, Ikeda M, et al. Cloning of a gene bearing missense mutations in early-onset familial Alzheimer's disease. *Nature (London)* 1995;**375**:754−60.
238. Wolozin B, Iwasaki K, Vito P, Ganjei JK, Lacana E, Sunderland T, et al. Participation of presenilin 2 in apoptosis: enhanced basal activity conferred by an Alzheimer mutation. *Science* 1996;**274**:1710−13.
239. Deng G, Pike CJ, Cotman CW. Alzheimer-associated presenilin-2 confers increased sensitivity to apoptosis in PC12 cells. *FEBS Lett* 1996;**397**:50−4.
240. Guo Q, Furukawa K, Sopher BL, Pham DG, Xie J, Robinson N, et al. Alzheimer's PS-1 mutation perturbs calcium homeostasis and sensitizes PC12 cells to death induced by amyloid β-peptide. *NeuroReport* 1996;**8**:379−83.
241. Guo G, Sopher BL, Pham DG, Furukawa K, Robinson N, Martin GM, et al. Alzheimer's presenilin mutation sensitizes neural cells to apoptosis induced by trophic factor withdrawal and amyloid β-peptide: involvement of calcium and oxyradicals. *J Neurosci* 1997;**17**:4212−22.
242. Kinoshita A, Whelan CM, Berezovska O, Hyman BT. Neurodegeneration induced by β-amyloid peptides in vitro: the role of peptide assembly state. *J Biol Chem* 2002;**277**:28530−6.
243. Ozaki T, Li Y, Kikuchi H, Tomita T, Iwatsubo T, Nakagawara A. The intracellular domain of the amyloid precursor protein (AICD) enhances the p53-mediated apoptosis. *Biochem Biophys Res Commun* 2006;**351**:57−63.
244. Alves da Costa C, Sunyach C, Parddossi-Piquard R, Sevalle J, Vincent B, Boyer N, et al. Presenilin-dependent γ-secretase-mediated control of p53- associated cell death in Alzheimer's disease. *J Neurosci* 2002;**26**:6377−85.
245. Esposito L, Gan L, Yu GQ, Essrich C, Mucke L. Intracellularly generated amyloid-β peptide counteracts the antiapoptotic function of its precursor protein and primes proapoptotic pathways for activation by other insults in neuroblastoma cells. *J Neurochem* 2004;**91**:1260−74.
246. Li WP, Chan WY, Lai HW, Yew DT. Terminal dUTP nick end labeling (TUNEL) positive cells in the different regions of the brain in normal aging and Alzheimer patients. *J Mol Neurosci* 1997;**8**:75−82.
247. Aoki M, Volkmann I, Tjernberg LO, Winblad B, Bogdanovic N. Amyloid β-peptide levels in laser capture microdissected cornu ammonis 1 pyramidal neurons of Alzheimer's brain. *NeuroReport* 2008;**19**(11):1085−9.

248. LaFerla FM, Tinkle BT, Bieberich CJ, Haudenschild CC, Jay G. The Alzheimer's Aβ peptide induces neurodegeneration and apoptotic cell death in transgenic mice. *Nat Genet* 1995;**9**:21−30.

249. Chui DH, Doboa E, Makifuchi T, Akiyama H, Kawakatsu S, Petit A, et al. Apoptotic neurons in Alzheimer's disease frequently show intracellular Aβ42 labeling. *J Alzheimer's Dis* 2001;**3**:231−9.

250. Nelson PT, Jicha GA, Schmitt FA, Liu H, Davis DG, Mendiondo MS, et al. Clinicopathologic correlations in a large Alzheimer Disease center autopsy cohort: neuritic plaques and neurofibrillary tangles 'do count' when staging disease severity. *J Neuropathol Exp Neurol* 2007;**66**:1136−46.

251. Kimura N, Yanagisawa K, Terao K, Ono F, Sakakibara I, Ishii Y, et al. Age-related changes of intracellular Aβ in cynomolgus monkey brains. *Neuropathol Appl Neurobiol* 2005;**31**:170−80.

252. Kakio A, Nishimoto SI, Yanagisawa K, Kozutsumi Y, Matsuzaka K. Cholesterol-dependent formation of GM1 ganglioside-bound amyloid-protein, an endogenous seed for Alzheimer amyloid. *J Biol Chem* 2001;**276**:24985−90.

253. Lord A, Kalimo H, Eckmang C, Zhang X-Q, Lannfelt L, Nilsson LGN. The Arctic Alzheimer mutation facilitates early intraneuronal Aβ aggregation and senile plaque formation in transgenic mice. *Neurobiol Aging* 2006;**27**:67−77.

254. D'Andrea MR, Nagele RG. Targeting the alpha 7 nicotinic acetylcholine receptor to reduce amyloid accumulation in Alzheimer's disease pyramidal neurons. *Curr Pharma Des* 2006;**12**:677−84.

255. Shoji M, Kawarabayashi T, Matsubara E, Ikeda M, Ishiguro K, Harigayay Y, et al. Distribution of amyloid β protein precursor in the AD brain. *Psychiatry Clin Neurosci* 2000;**54**:45−54.

Intraneuronal Amyloid and Plaque Formation

A myloid plaques are one of the most studied neuropathological features of Alzheimer's disease (AD), and yet after all these years, they still remain an enigma. This chapter will present several points-of-view about how plaques form, how they are characterized, their relationship of intraneuronal Aβ, as well as their relationship to NFTs, the other neuropathological feature of AD.

MODELS OF PLAQUE FORMATION

The widely accepted model of amyloid plaque formation through neuronally secreted Aβ is the basis of the Amyloid hypothesis. Recently,

Intracellular Consequences of Amyloid in Alzheimer's Disease.
DOI: http://dx.doi.org/10.1016/B978-0-12-804256-4.00005-X

TABLE 5.1   Comparing the 3 Hypothetical Models of Amyloid-Based Models of Plaque Formation

Measures	Amyloid cascade hypothesis[1] (deposition-based)	Synaptic hypothesis[2–4] (deposition-based)*		Lytic Model[5] (lysis-based)
Intraneuronal source of plaque Aβ	Yes	Yes	Yes	Yes
Intraneuronal uptake of Aβ	Not reported	Not reported	Yes	Yes
Neurons secrete Aβ (deposition)	Yes	Yes	Not clear	No**
Location of plaques	Near neurons	In synapses	In neurites	Is the neuron**

*Two variations presented.
**Senile, dense-core amyloid plaques only.

a revised model was proposed, known as the Synaptic hypothesis. This model proposed that amyloid oligomers accumulate in the synapses, leading to synaptic failure, neurodegeneration, and AD. Alternatively, about 15 years ago, "Inside-Out" hypothesis (presently renamed the Lytic Model) was published, which presented a distinct model: neurons overaccumulate (not overproduce) Aβ resulting in their death as the dense-core, senile amyloid plaque. Each of these hypotheses is compared in Table 5.1 and can be generally classified as deposition- or lytic-based models.[1–5]

Deposition-Based Models

Random Deposition Into the Parenchyma

The Amyloid hypothesis taught the AD research community that the amyloid plaque is the quintessential cornerstone of the AD brain that forms from the secretions of Aβ from the amyloid overproducing neurons. As the neurons secrete the Aβ as deposits, it aggregates to form the toxic senile plaques that kill neurons. Most, if not all diagrams of AD neuropathology show this process.[6] On perhaps one of the most viewed AD websites, it is very clear that plaques composed of chemically "sticky" clusters of protein, build up or aggregate BETWEEN neurons. The identification of Aβ as the major component of the senile plaque leads to this hypothesis.[7] Fig. 5.1 demonstrates this belief that neurons secrete amyloid that aggregate to form the extracellular senile amyloid plaques.

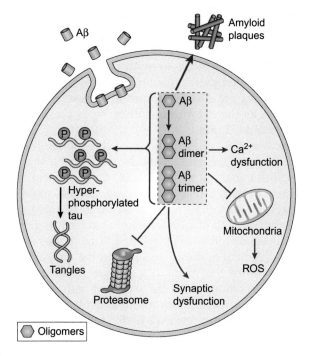

FIGURE 5.1 Pathological effects of intraneuronal Aβ. Amyloid-β (Aβ), produced intracellularly or taken up from extracellular sources, has various pathological effects on cell and organelle function. Intracellular Aβ can exist as a monomeric form that further aggregates into oligomers, and it may be any of these species that mediate pathological events in vivo, particularly within a dysfunctional neuron. Evidence suggests that intracellular Aβ may contribute to pathology by facilitating tau hyperphosphorylation, disrupting proteasome and mitochondria function, and triggering calcium and synaptic dysfunction. *Aβ*, amyloid-β; *Ca*, calcium; *ROS*, reactive oxygen species. *Source: Used with permission from* Nat Rev Neurosci *2007;8:499−509.*

The origins of this model of plaque formation may have originated in the 1970s. Initially microglia were thought to be the source of the plaque amyloid since they were localized at very early stages of neuritic plaque formation. Subsequent data showed that the cells were neurons leading investigators to propose that the Aβ in senile plaques is derived from nerve cells.[8] Additional data confirmed that the aggregated amyloid peptides in plaques are derived from secretory processing of the APP in the neurons.[9]

The aggregated Aβ is composed of a short peptide of 39−42 amino acids in length, which is produced by the sequential cleavage of APP. Of the forms of the cleaved APP products, three different isoforms exist: the shortest isoform, APP695, is mostly expressed by neurons, whereas the two longer isoforms, APP751 and APP770, are expressed predominantly in glial cells, such as astrocytes.[10–14] Another report indicated that the plaques begin with the aggregation of the longer Aβ42 peptides that are associated with dystrophic neurites, reactive microglia, and activated astrocytes.[13,14] Regardless of the form of Aβ, extracellular plaques are observed, and are associated with neuronal loss in areas such as the entorhinal cortex, hippocampus, and association cortices in the AD brain.[15]

The preclinical models of AD reinforced this model. It was proposed that the seeding or aggregation may be due to the inability to hide hydrophobic C-terminal parts of Aβ in the intraneuronal micelles, perhaps because the structure grows too large and becomes unstable.[16]

Others suggest that glial cells contribute to plaque formation. Astrocytes might promote plaque formation and maturation.[17] The high level of α7 expression on astrocytes especially in the APPSwe brain, suggests an involvement of the α7 receptor in the metabolism of APP as well as with the formation of neuritic plaques.[18] Several investigators propose that the secretion of Aβ by activated astrocytes could significantly contribute to the formation of amyloid plaques in AD.[19–21] The microglia, like that of astrocytes, may under some circumstances exacerbate inflammatory processes, and induce additional neuronal damage while advancing plaque formation.[22]

Synaptic Deposition

The Synaptic hypothesis, a variation of the initial deposition model, pinpoints the area of the neuronal-secreted amyloid to the synapse suggesting that the extracellular release of Aβ from degenerating neurites might upregulate Aβ42 within adjacent synaptic compartments, leading to the synaptic spread of AD (Fig. 5.2, left panel).[2,23] The accumulation of APP in synaptosomes and/or synaptic plasma membranes may lead to Aβ overproduction in these compartments, and the accumulated intracellular Aβ in turn, may be secreted extracellularly to accelerate senile plaque formation. Therefore, Aβ secreted extracellularly accelerates plaque formation. The nerve ending may be where senile plaques are initially formed or may be significantly associated with spontaneous plaque formation that occurs with advancing age.[24,25]

A modification of this synaptic model suggests that intracellular Aβ accumulates in the distal neuronal processes leading to synaptic degeneration as the amyloid plaque in-place (Fig. 5.2, right panel). Most importantly, as the Aβ42 accumulates progressively in distal processes after the initial increases, the Aβ42 declines in cell bodies as the plaques develop preferentially in distal processes and synapses.[26,27] In support,

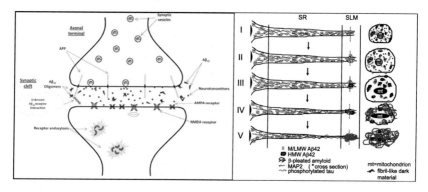

FIGURE 5.2   Left panel. Soluble Aβ oligomers promote receptor endocytosis, reducing the density of the receptors at the synapses. Aβ is secreted into the synaptic cleft via sequential cleavage of presynaptic APP (internally or at the cell surface) by β-secretase and γ-secretase or gains entry from outside the synapse. The accumulation of Aβ oligomers in the synaptic cleft leads to reduced NMDA and AMPA receptor density in synapses, leading to attenuated LTP and neurotransmission. While Aβ oligomers may play a normal role in controlling LTP, accelerated synaptic accumulation of Aβ oligomers (eg, due to FAD gene mutations) may lead to a toxic gain of function and cognitive decline. Right panel. Schematic diagram of proposed sequence of Aβ42, MAP2, and tau alterations in a CA1 pyramidal cell apical dendrite with aging. At top (I), normal dendrite contains MAP2 associated with microtubules and M/LMW Aβ42 peptides. Cross section taken through the distal apical dendrite (SLM) is shown at right. (II) With aging, M/LMW Aβ42 peptides accumulate, which coincides with early reductions in MAP2, especially in the SLM. Later on in Tg2576 mice, Aβ42 M/LMW peptides co-localize with hyperphosphorylated tau in distal processes and synaptic compartments; this co-localization is more prominent in 36 Tg mice (23). (III) Subsequently, Aβ42 HMW oligomers develop in the distal dendrite, which is associated with localized absence of MAP2. Concomitantly, Aβ42 M/LMW peptides further accumulate in more proximal regions of the dendrite. (IV) This is followed by Aβ fibril formation, especially in distal neurites of the SLM and (V) deposition of amyloid plaques in SLM. *Aβ42*, amyloid-β42; *AMPA*, α-amino-3-hydroxy-5-methyl-4-isoxazolepropionic acid; *FAD*, familial Alzheimer's disease; *HMW*, high molecular weight; *LMW*, low molecular weight; *LTP*, long-term potentiation; *MAP2*, microtubule-associated protein 2; *M/LMW*, monomer and low molecular weight; *mt*, mitochondrion; *NMDA*, N-methyl-D-aspartate; *SLM*, stratum lacunosum-moleculare; *Tg2576*, Swedish mutant transgenic mice. *Source: Left panel: Used with permission from* Biol Rep 2009;1:54.*; right panel: Courtesy of* PLoS ONE 2013;8(1):e51965.

the amount of intracellular Aβ increased in the cellular fractions of nerve ending with age, and intraneuronal levels of Aβ42 decline with the emergence of extracellular Aβ plaques.[27,28]

Accumulated pools of intraneuronal Aβ were detected in early AD and in young Down syndrome brains that declined with age and plaque burden in Down syndrome brains.[5,25–32] Intraneuronal Aβ peaked around the onset of extracellular amyloid deposition in transgenic mice, and later declined with increasing extracellular plaque density suggesting a dynamic equilibrium between intra- and extracellular Aβ deposition.[33] Therefore, the drainage of intraneuronal Aβ by senile plaques likely explains why intracellular Aβ has been rarely described in AD postmortem brain with abundant senile plaque deposition.[16]

Lytic Model

In 1999, well before the release of the disappointing results from the amyloid-based clinical trials, a novel lysis-based model was proposed (independent of Aβ deposition and neuronal overproduction) whereby plaques are the result of the Aβ-overburdened lysed neurons (see Chapter 9).[6] The pathological events leading to the lysis of the neuron begin with the entry of Aβ into the brain through a dysfunctional BBB.[34] Copious amounts of previously CNS-regulated Aβ bind with high affinity to the α7 neuronal receptors, and endocytose until the neurons are overburdened, and then die. As the neurons die, the once-contained lytic enzymes are released along with all other neuronal components in the brain leading to the death of additional neurons through the secondary consequences of inflammation, as well as through the destruction of their processes that passed through this highly toxic area (Fig. 5.3).[6]

Others hypothesized that intraneuronal Aβ aggregates are micelles of self-associated Aβ peptides that later serve as nuclei for Aβ fibrillization once released by neurons or following "cell lysis" supporting the Lytic Model.[6,16] One study reported no dramatic changes in the distribution of APP mRNA in conjunction with the development of senile plaques or NFTs in AD.[35] Not only does this further emphasize the impractical possibility that the neuron contributes its Aβ to build plaques, but the excessive amounts of intraneuronal Aβ must originate from outside the neuron.

Aβ is localized in lysosomes, as well as endosomes and multivesicular bodies.[36–40] Co-localization of Aβ42 with the lysosomal marker LAMP-1 further confirms this possibility.[41] The acidic pH in lysosomes provides favorable conditions for the concentrated amounts of intralysosomal Aβ aggregate or oligomerize.[36,42–45] Coupled with the presence of early abnormalities in the endosomal-lysosomal system that precedes amyloid deposition in sporadic AD and Down syndrome, and since lysosomes are typically located in the cell body, it is only logical to expect preferential intraneuronal accumulation of Aβ in the cell body (perikaryon).[38,46,47] Also, in the mLRP-transfected mice, the endocytosis of extracellular Aβ is not degraded completely and so accumulates in the neurons.[41]

Aggregates of intraneuronal Aβ in the brains of the APP transgenic mice did not stain with thioflavin S, a stain of fibrillar Aβ, suggesting

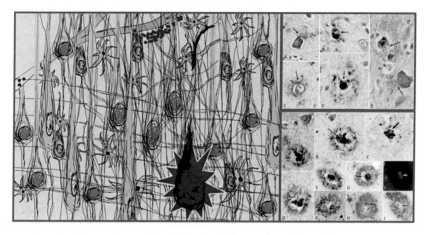

FIGURE 5.3   Left panel. The Lytic Model. Right top panel. Representative examples of Aβ42 immunolabeling in AD cortical neurons show early stages of cell lysis (*arrows*). Right bottom panel. Representative examples of Aβ42 dense-core (senile/neuritic) plaques (labeled brown in panels A–E, J; labeled black in panel F; labeled red in panels G–H) in AD cortical tissue sections. Note the hematoxylin-stained, purple nuclei (*arrows*) deeply embedded in the middle of the plaques (B–H). Double IHC using a detection system to label Aβ42 black, and to label NeuN (*arrow*), red, thereby confirming the presence of a neuronal nucleus at the center of this amyloid plaque (F). Blue DAPI-dyed nuclei (I, *arrow*) is observed in the center of the amyloid plaque (J). *Aβ42*, amyloid-β42; *NeuN*, neuronal-specific nuclear protein. *Source: Drawing used with permission from Slidomics, LLC; IHC images used with permission from* Histotech Biochem *2010;85(5):133–47;* Curr Pharm Dis *2006;12(6):677–84;* Histopathology *2001;38:120–34.*

that the additional lytic activity must occur as the neurons die for the released amyloid to become thioflavin S-positive. Previous studies support this possibility since active enzymatic cathepsin D activity has been detected in plaques, making the lysed neuron a credible source. Furthermore, intracellular injection of Aβ peptides leads to cell death of cultured human cortical brain neurons.[7,48]

This Lytic Model can readily explain the presence of neuronal DNA in plaques, which cannot be explained by the deposition-based models.[6,49] The model can explain the presence of other neuronal components in plaques such as lipofuscin and cathepsin D, neuronal ribonucleic acid (RNA), and ubiquitin that are typically located in the perikaryon, or cell body.[6,50] This Lytic Model also explains the presence of amyloid plaques in the pyramidal cell layers that correlate with sizes of pyramidal cell bodies.

If the synaptic models are accurate, then plaques should form more readily in the molecular layer of the brain, areas filled with synapses and devoid of neuronal cell bodies, and yet, such amyloid plaques are rarely detected in that area. The deposition-based models cannot readily explain the typical spherically-shaped amyloid plaques as well.

INTRANEURONAL Aβ THEN PLAQUES

Several clinical manifestations of AD cannot be entirely explained by the traditional amyloid hypothesis. There is no correlation between the number, location, and distribution of amyloid plaques and parameters of AD neuropathology, including the degree of dementia and neurodegeneration.[51] Perhaps the lone variable that seems to support the amyloid hypothesis is that soluble Aβ concentrations in brain are highly correlated with severity of disease in AD leading some to suggest that soluble Aβ may cause initial neuritic changes and NFTs.[52−54] Others believe that soluble Aβ is not toxic enough to produce such widespread neuronal death. Despite the fact that plaque-associated Aβ also increases with age, the spatial distribution did not match with the pattern of neuron loss. A possible toxic effect of soluble Aβ species is likely to be negligible because other groups of neurons like dentate gyrus granule cells, which do not accumulate intraneuronal Aβ, are unaffected.[55]

Reports are repeatedly finding that levels of intraneuronal Aβ that increase with age, occur before the appearance of extracellular amyloid plaques in patients with Down syndrome and AD as well as in many preclinical models of AD.[7,16,44,56−63] Biochemical data confirmed abundant intraneuronal Aβ42 and Aβ40 well before the appearance of extracellular plaques.[58] This notion of an intracellular pathology preceding extracellular plaque deposition is also supported by an independent immunocytochemical study showing an age-dependent accumulation of intraneuronal Aβ42 in non-AD subjects, and a preferential accumulation of intraneuronal Aβ42 in AD-vulnerable versus AD-resistant brain regions.[26] Intraneuronal Aβ deposition in the somas of hippocampal CA1/subiculum neurons was observed far in advance of the occurrence of extracellular Aβ plaques.[64]

CHARACTERIZING PLAQUE TYPES: DOES IT MATTER?

Simply stated yes. Several report the lack of correlation between plaque load and cognitive impairment, but there is little definition of the kinds of plaques that are being referenced. Not all of the amyloid plaques detected in the AD brain are the same, whereby some are without inflammatory or cell death consequences, as others are the result of cell deaths and trigger gliosis (see Chapter 7).[6,65] This differential feature may explain how these plaques form: dense-core as the result of the lysed, Aβ42-overburdened neuron, and diffuse plaques form not from the reported mechanisms of deposition, but by simple leakage from dysfunctional vessels (Fig. 5.4).

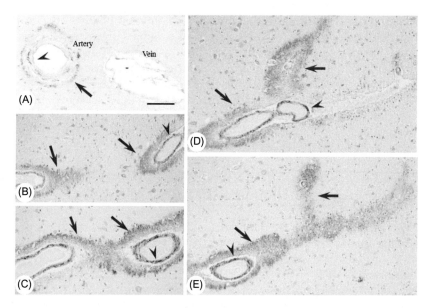

FIGURE 5.4 Aβ42-positive, vascular, diffuse plaques associated with an artery (*arrow*) and but not with a vein (A). Vascular-associated Aβ42 resembling diffuse plaques (*arrows*) in serial sectioning sets (B and C, D and E) clearly show its association with the vessels (panels B–E). Also note detection of intracellular Aβ42 in vascular smooth muscle cells (*arrowheads*). *Aβ42*, amyloid-β42. *Source: Used with permission from* Biotech Histochem *2010;85(5):133–47.*

There is nothing random about the morphology of the extracellular Aβ, which neither deposition-based hypothetical model can explain (Fig. 5.4). The extracellular Aβ is obviously associated with the vessels (arteries not veins) thereby providing visual evidence of a leaky vascular system while defining the existence of a unique type of plaque; a diffusely stained plaque that will not evolve into a senile, dense-core plaque.[6]

Perhaps the source of this ambiguity lies in the early teachings of the amyloid cascade and revised synaptic hypotheses that all plaques are created equally through deposition. Early names of plaques were based on the morphology of their staining characteristics, and the name-calling was exhaustive.[6] However, while the most frequent names were senile and neuritic, more recent names used to describe amyloid plaques are diffuse and dense-core, which are based on their immuno-histochemical labeling morphology. The widely-held belief is that the diffuse amyloid plaques represent an immature senile plaque. This hypothesis was challenged in 2002 stating that these two plaque types do not represent the same plaque over time. Therefore, diffuse plaques will never mature or evolve into dense-core plaques.[66]

Contrary to the popularized dogma that all amyloid plaques arise from extracellular deposition, plaques may also originate from vessels, neurons, Purkinje cell dendritic processes, and astrocytes.[67] Another group hypothesized that CAA may be a seeding origin of the dense-core plaques because they observed blood vessels in the middle of such plaques.[68]

It is important to differentiate the type of plaque as some plaque types are associated with cognitive decline and inflammation while others are not.[6]

RELATIONSHIP OF INTRANEURONAL Aβ, PLAQUES, AND NEUROFIBRILLARY TANGLES

Plaques and tangles are the neuropathological features of AD neuropathology, and now intraneuronal Aβ is among these hallmarks. Intraneuronal Aβ is detected before plaques, and extracellular deposits of Aβ precede NFTs in the AD, Down syndrome, and in transgenic mice brains.[7,69] Amyloid plaques were detected in a middle-aged AD patient with the FAD mutation before the presence of NFTs. Tau mutants can cause fontotemporal dementia where many tangles, but not amyloid deposits, are found in the human brain suggesting that the formation of tangles and plaques is independent. Similar results were reported in the APP/PS1KI mice where toxic Aβ peptides accumulate intra- and extracellularly and coincided with hippocampus atrophy independent of neurofibrillary tangle formation.[55]

One of the explanations for co-localization of amyloid plaques with NFTs is based on their interaction. The stretch of amino acids in Aβ (25−35) is able to aggregate into fibrils, while the residues 307−325 in tau involve microtubule binding.[73] Their interaction can produce the aggregation of tau into NFTs, and the concomitant dissembling of Aβ25-35 suggesting that the tau protein may have a protective action by preventing Aβ from adopting the cytotoxic, aggregated form.[70]

A potential mechanism of Aβ-induced tau pathology suggested that "Aβ as the accelerator or initiator, and tau as the executor."[71] Extracellular injection of synthetic pre-aggregated Aβ peptides in the brains of tau transgenic mice resulted in NFTs in neurons remote from the injection site but connected with the site of injection.[72] This indicates that aggregated amyloid peptides can initiate and propagate tau aggregation in functionally connected brain regions.[72,73] Tau pathology did not induce amyloid pathology in any in vivo studies,[71] however; it is not clear how amyloid pathology was defined.

In vitro findings using neuroblastoma and ex vivo synaptosomes demonstrated increased tau phosphorylation (Ser-202, Thr-181, and Thr-231) following Aβ peptide binding to α7 receptor suggesting Aβ receptor-mediated tau pathology.[71] In personal communication, all neurons with intraneuronal NFTs were positive for Aβ42 and α7 receptors (Fig. 5.5). Interestingly, as far back as 1912, Oskar Fisher did not observe tangles without plaques suggesting that the tangle formation was subsequent to detectable amyloid pathology.[74]

The accumulation of intraneuronal Aβ occurred concomitantly with the hyperphosphorylation of tau within synaptic terminals in the triple-transgenic mice.[24] In the transgenic rat model, intraneuronal accumulation of Aβ upregulated the phosphorylated form of extracellular-regulated kinase-2 and its enzymatic activity, which led to increased levels of tau phosphorylation that correlated with a mild spatial learning deficit.[56]

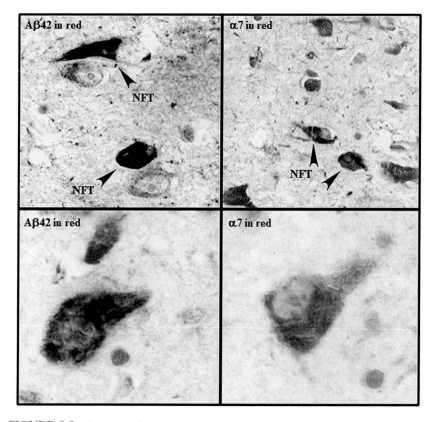

FIGURE 5.5 Presence of NFTs (stained black) in Aβ42 (left panels)- and α7 receptor (right panels)-positive neurons in the Alzheimer's brain at low (60x objective; top panels) and high (100x objective; bottom panels) magnifications. Areas of Aβ42-positive plaques were present in the AD tissues but are not presented. α7 receptor, alpha 7 nicotinic acetylcholine receptor; Aβ42, amyloid-β42; NFT, neurofibrillary tangles. *Source: Used with permission from Slidomics, LLC.*

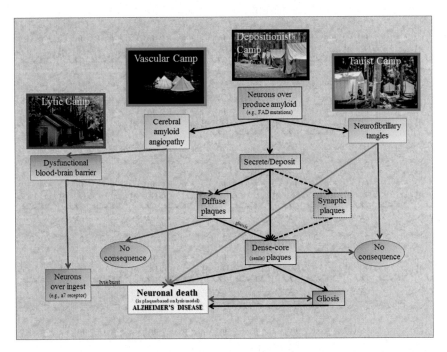

FIGURE 5.6   Alzhemier's disease models. The relationship between four hypothetical models (camps) of the causes of Alzhiemer's diease are presented: the Lytic, Vascular, Depositionist (Baptist), and Tauist camps. The original Baptist Model, or the amyloid cascade (deposition) hypothesis, is based on the overproduction of Aβ by neurons that is secreted or deposited into the extracellular space in the brain (or synapses [*dotted block lines*]), which may initially form the toxic diffuse amyloid plaques that mature overtime with the possible help of the inflammatory cells to form the dense-core, senile plaques that kill nearby neurons either directly, or indirectly through the process of gliosis that lead to neuronal death (*solid black lines*). The Tauist Model simply states that the formation of hyperphosphorylated tau as neurofibrillary tangles in neurons leads to neuronal cell death then to AD (*blue line*). The Vascular-based Model considers the pathological consequences of the vascular system (further discussed in Chapter 8) as the primary cause of neuronal death, and therefore, Alzheimer's disease (*green line*). The Lytic Model is based on the overaccumulation of intraneuronal Aβ from peripheral sources that gain unregulated entry into the CNS through a dysfunctional blood-brain barrier. Overaccumulation in the neurons leads to cell lysis (neuronal death as the dense-core, senile plaque) and to gliosis that contributes to additional neuronal death independent of amyloid. The Lytic Model also shows that (1) diffuse amyloid plaques originate from leaky vessels and are without any pathological consequences; meaning they are not toxic, and will never evolve into dense-core, senile plaques, (2) reported senile plaques that evolve from the so-called diffuse plaques are also without consequence to cognitive impairment unless they are the result of neuronal death, and (3) neurofibrillary tangles are consequences, and not the primary causes of neuronal degeneration (*red lines*). α7 receptor, alpha 7 nicotinic acetylcholine receptor; Aβ42, amyloid-β42; CNS, central nervous system. *Source: Used with permission from Slidomics, LLC.*

SUMMARY

The information provided in this chapter is the basis of several hypothetical models of the causes of AD. Before the interest in intracellular Aβ, AD research seemed to be divided into three lines of research: those who believe Aβ was the cause of AD (Baptist), those who believe tau was the cause of the disease (Tauists), and those who believe that the vascular pathology is the root cause of AD.

Now that investigations of the potential toxic role of intraneuronal Aβ are becoming more and more investigated, new hypotheses have emerged. The Synaptic plaque hypothesis, like the Baptist amyloid cascade hypothesis, is based on the deposition of Aβ in the synaptic terminals, or in the synaptic space. The Lytic Model is based on the overaccumulation of Aβ into the neurons that lyse forming the dense-core, senile plaques triggering additional neuronal death independent of intraneuronal Aβ (Fig. 5.6).

References

1. Hardy J, Selkoe DJ. Review. The amyloid hypothesis of Alzheimer's disease: progress and problems on the road to therapeutics. *Science* 2002;**297**:353−6.
2. Hooli BV, Tanzi RE. Review. A current view of Alzheimer's disease. *Biol Rep* 2009;**1**:54.
3. Takahashi RH, Capetillo-Zarate E, Lin MT, Milner TA, Gouras GK. Accumulation of intraneuronal β-amyloid 42 peptides is associated with early changes in microtubule-associated protein 2 in neurites and synapses. *PLoS ONE* 2013;**8**(1):e51965.
4. Tampellini D, Gouras GK. Review. Synapses, synaptic activity and intraneuronal Aβ in Alzheimer's disease. *Front Aging Neurosci* 2010;**2**(13):1.
5. D'Andrea MR. *Bursting neurons and fading memories*. New York: Elsevier Press; 2014.
6. <http://www.alz.org/norcal/in_my_community_20545.asp>: Alzheimer's Basics; Plaques and Tangles.
7. Xiaqin S, Yan Z. Amyloid Hypothesis and Alzheimer's Disease. In: Chang R C-C, editor. *Advanced Understanding of Neurodegenerative Diseases*. InTech; 2011. ISBN: 978-953-307-529-7, Available from: < http://www.intechopen.com/books/advanced-understanding-of-neurodegenerativediseases/amyloid-hypothesis-and-alzheimer-s-disease>.
8. Goedert M. Neuronal localization of amyloid beta protein precursor mRNA in normal human brain and in AD. *EMBO J* 1987;**6**(12):3627−32.
9. Young MJ, Lee RK, Jhaveri S, Wurtman RJ. Intracellular and cell-surface distribution of amyloid precursor protein in cortical astrocytes. *Brain Res Bull* 1999;**50**(1):27−32.
10. Chow VW, Mattson MP, Wong PC, et al. Review. An overview of APP processing enzymes and products. *Neuromolecular Med* 2010;**12**:1−12.
11. LeBlanc AC, Chen HY, Autilio-Gambetti L, et al. Differential APP gene expression in rat cerebral cortex, meninges, and primary astroglial, microglial and neuronal cultures. *FEBS Lett* 1991;**292**:171−8.
12. Mills J, Reiner PD. Regulation of amyloid precursor protein cleavage. *J Neurochem* 1999;**72**:443−60.
13. Murphy MP, Levine H. Alzheimer's disease and the amyloid-β peptide. *J Alzheimers Dis* 2010;**19**:311−23.

14. Selkoe DJ, Schenk D. Alzheimer's disease: molecular understanding predicts amyloid-based therapeutics. *Annu Rev Pharmacol Toxicol* 2003;**43**:545–84.
15. Braak H, Braak E. Evolution of neuronal changes in the course of Alzheimer's disease. *J Neural Transm Suppl* 1998;**53**:127–40.
16. Lord A, Kalimo H, Eckmang C, Zhang X-Q, Lannfelt L, Nilsson LGN. The Arctic Alzheimer mutation facilitates early intraneuronal Aβ aggregation and senile plaque formation in transgenic mice. *Neurobiol Aging* 2006;**27**:67–77.
17. Terai K, Iwai A, Kawabata S, Sasamata M, Miyata K, Yamaguchi T. Apolipoprotein E deposition and astrogliosis are associated with maturation of beta-amyloid plaques in beta APPswe transgenic mouse: implications for the pathogenesis of Alzheimer's disease. *Brain Res* 2001;**900**:48–56.
18. Yua W-F, Guana Z-Z, Bogdanovic N, Nordberg A. High selective expression of α7 nicotinic receptors on astrocytes in the brains of patients with sporadic Alzheimer's disease and patients carrying Swedish APP 670/671 mutation: a possible association with neuritic plaques. *Exp Neurol* 2005;**192**:215–25.
19. Busciglio J, Gabuzda BH, Matsudaira P, Yanker BA. Generation of β-amyloid in the secretory pathway in neuronal and nonneuronal cells. *PNAS* 1993;**90**:2092–6.
20. Itagaki S, McGeer P, Akiyama H, Zhu S, Selkoe D. Relationship of microglia and astrocytes to amyloid plaques of Alzheimer disease. *J Neuroimmunol* 1989;**24**:173–83.
21. Siman R, Card J, Nelson R, Davis L. Expression of β-amyloid precursor protein in reactive astrocytes following neuronal damage. *Neuron* 1989;**3**:275–85.
22. Shoham S, Ebstein RP. The distribution of β-amyloid precursor protein in rat cortex after systemic kainite-induced seizures. *Exp Neurol* 1997;**147**:361–76.
23. Takahashi RH, Capetillo-Zarate E, Lin MT, Milner TA, Gouras GK. Co-occurrence of Alzheimer's disease β-amyloid and tau pathologies at synapses. *Neurobiol Aging* 2010;**31**(7):1145–52.
24. Kimura N, Nakamura S, Honda T, Takashima A, Nakayama H, et al. Age-related changes in the localization of presenilin-1 in cynomolgus monkey brain. *Brain Res* 2001;**922**:30–41.
25. Kimura N, Tanemura K, Nakamura S, Takashima A, Ono F, Sakakibara I, et al. Age-related changes of Alzheimer's disease-associated proteins in cynomolgus monkey brains. *Biochem Biophys Res Commun* 2003;**310**:303–11.
26. Gouras GK, Tsai J, Näslund J, Vincent B, Edgar M, Checler F, et al. Intraneuronal Aβ42 accumulation in human.. *Brain Am J Pathol* 2000;**156**(1):15–20.
27. LaFerla FM, Green KN, Oddo S. Review. Intracellular amyloid-β in Alzheimer's disease. *Nat Rev Neurosci* 2007;**8**:499–509.
28. Kimura N, Yanagisawa K, Terao K, Ono F, Sakakibara I, Ishii Y, et al. Age-related changes of intracellular Aβ in cynomolgus monkey brains. *Neuropathol Appl Neurobiol* 2005;**31**:170–80.
29. D'Andrea MR, Nagele RG, Wang HY, Peterson PA, Lee DH. Evidence that neurones accumulating amyloid can undergo lysis to form amyloid plaques in Alzheimer's disease. *Histopathology* 2001;**38**:120–34.
30. Gyure KA, Durham R, Stewart WF, Smialek JE, Troncoso JC. Intraneuronal abeta-amyloid precedes development of amyloid plaques in Down syndrome. *Arch Pathol Lab Med* 2001;**125**:489–92.
31. Mori C, Spooner ET, Wisniewsk KE, Wisniewski TM, Yamaguchi H, Saido TC, et al. Intraneuronal Aβ42 accumulation in Down syndrome brain. *Amyloid* 2002;**9**: 88–102.
32. Tu S, Okamoto S-I, Lipton SA, Xu H. Oligomeric Aβ-induced synaptic dysfunction in Alzheimer's disease. *Mol Neurodegener* 2014;**9**:48.

33. Oddo S, Billings L, Kesslak JP, Cribbs DH, LaFerla FM. Aβ immunotherapy leads to clearance of early, but not late, hyperphosphorylated tau aggregates via the proteasome. *Neuron* 2004;**43**:321−32.

34. <https://www.youtube.com/watch?v = _NTaGjQow1c> [accessed 07.10.15].

35. Hyman BT, Wenniger JJ, Tanzi RE. Nonisotopic in situ hybridization of amyloid β protein precursor in Alzheimer's disease: expression in neurofibrillary tangle bearing neurons and in the microenvironment surrounding senile plaques. *Mol Brain Res* 1993;**18**:253−8.

36. Takahashi RH, Almeida CG, Kearney PF, Yu F, Lin MT, Milner TA, et al. Oligomerization of Alzheimer's β-amyloid within processes and synapses of cultured neurons and brain. *J Neurosci* 2004;**24**:3592−9.

37. Shie FS, LeBoeuf RC, Jin LW. Early intraneuronal Aβ deposition in the hippocampus of APP transgenic mice. *Neuroreport* 2003;**14**:123−9.

38. Cataldo AM, Petanceska S, Terio NB, Peterhoff CM, Durham R, Mercken M, et al. Aβ localization in abnormal endosomes: association with earliest Aβ elevations in AD and Down syndrome. *Neurobiol Aging* 2004;**25**:1263−72.

39. Takahashi RH, Milner TA, Li F, Nam EE, Edgar MA, Yamaguchi H, et al. Intraneuronal Alzheimer Aβ42 accumulates in multivesicular bodies and is associated with synaptic pathology. *Am J Pathol* 2002;**161**:1869−79.

40. Langui D, Girardot N, El Hachimi KH, Allinquant B, Blanchard V, Pradier L, et al. Subcellular topography of neuronal Aβ peptide in APPxPS1 transgenic mice. *Am J Pathol* 2004;**165**:1465−77.

41. Zerbinatti CV, Wahrle SE, Kim H, Cam JA, Bales K, Paul SM, et al. Apolipoprotein E and low density lipoprotein receptor-related protein facilitate intraneuronal Aβ42 accumulation in amyloid model mice. *J Biol Chem* 2006;**281**(47):36180−6.

42. Stine Jr WB, Dahlgren KN, Krafft GA, LaDu MJ. Subcellular topography of neuronal Aβ peptide in APPxPS1 transgenic mice. *J Biol Chem* 2003;**278**:11612−22.

43. Pasternak SH, Callahan JW, Mahuran DJ. The role of the endosomal/lysosomal system in amyloid-β production and the pathophysiology of Alzheimer's disease: reexamining the spatial paradox from a lysosomal perspective. *J Alzheimers Dis* 2004;**6**:53−65.

44. Leon WC, Canneva F, Partridge V, Allard S, Ferretti MT, DeWilde A, et al. A novel transgenic rat model with a full Alzheimer's-like amyloid pathology displays pre-plaque intracellular amyloid-β-associated cognitive impairment. *J Alzheimers Dis* 2010;**20**:113−26.

45. Walsh DM, Tseng BP, Rydel RE, Podlisny MB, Selkoe DJ. The oligomerization of amyloid β-protein begins intracellularly in cells derived from human brain. *Biochemistry* 2000;**39**:10831−9.

46. Cataldo AM, Barnett JL, Pieroni C, Nixon RA. Increased neuronal endocytosis and protease delivery to early endosomes in sporadic Alzheimer's disease: neuropathologic evidence for a mechanism of increased β-amyloidogenesis. *J Neurosci* 1997;**17**:6142−51.

47. Roberts VJ, Gorenstein C. Examination of the transient distribution of lysosomes in neurons of developing rat brains. *Dev Neurosci* 1987;**9**(4):255−64.

48. Echeverria V, Cuello AC. Review. Intracellular Aβ amyloid, a sign for worse things to come? *Mol Neurobiol* 2002;**26**(2−3):299−316.

49. D'Andrea MR, Reiser PA, Gumula NA, Hertzog BM, Andrade-Gordon R. Application of triple immunohistochemistry to characterize amyloid plaque-associated inflammation in brains with Alzheimer's disease. *Biotech Histochem* 2001;**76**(2):97−106.

50. D'Andrea MR, Nagele RG, Gumula NA, Reiser PA, Polkovitch DA, Hertzog BM, et al. Lipofuscin and Aβ42 exhibit distinct distribution patterns in normal and Alzheimer's disease brains. *Neurosci Lett* 2002;**323**:45−9.

51. Swaab DF, Lucassen PJ, van de Nes AP, Ravid R, Salehi A. In: Hyman BT, Duyckaerts C, Christen Y, editors. *Connections, Cognition, and Alzheimer's Disease.* New York: Springer-Verlag New York Inc.; 1997. p. 83−104.

52. Lue LF, Kuo YM, Roher AE, Brachova L, Shen Y, Sue L, et al. Soluble amyloid β peptide concentration as a predictor of synaptic change in Alzheimer's disease. *Am J Pathol* 1999;**155**:853−86.

53. McLean CA, Cherny RA, Fraser FW, Fuller SJ, Smith MJ, Beyreuther K, et al. Soluble pool of Aβ amyloid as a determinant of severity of neurodegeneration in Alzheimer's disease. *Ann Neurol* 1999;**46**:860−6.

54. Näslund J, Haroutunian V, Mohs R, Davis KL, Davies P, Greengard P, et al. Correlation between elevated levels of amyloid β-peptide in the brain and cognitive decline. *JAMA* 2000;**283**:1571−7.

55. Breyhan H, Wirths O, Duan K, Marcello A, Rettig J, Bayer TA. APP/PS1KI bigenic mice develop early synaptic deficits and hippocampus atrophy. *Acta Neuropathol* 2009;**117**:677−85.

56. Echeverria V, Ducatenzeiler A, Dowd E, Jänne J, Grant SM, Szyf M, et al. Altered mitogen-activated protein kinase signaling, tau hyperphosphorylation and mild spatial learning dysfunction in transgenic rats expressing the β-amyloid peptide intracellularly in hippocampal and cortical neurons. *Neuroscience* 2004;**129**(3):583−92.

57. D'Andrea MR, Nagele RG. Targeting the alpha 7 nicotinic acetylcholine receptor to reduce amyloid accumulation in Alzheimer's disease pyramidal neurons. *Curr Pharm Des* 2006;**12**:677−84.

58. Iulita MF, Allard S, Richter L, Munter L-M, Ducatenzeiler A, Weise C, et al. Intracellular Aβ pathology and early cognitive impairments in a transgenic rat overexpressing human amyloid precursor protein: a multidimensional study. *Acta Neuropathol Commun* 2014;**2**:61.

59. Wirths O, Multhaup G, Czech C, Feldmann N, Blanchard V, Tremp G, et al. Intraneuronal APP/Aβ trafficking and plaque formation in β-amyloid precursor protein and presenilin-1 transgenic mice. *Brain Pathol* 2002;**12**:275−86.

60. Ferretti MT, Partridge V, Leon WC, Canneva F, Allard S, Arvanitis DN, et al. Transgenic mice as a model of pre-clinical Alzheimer's disease. *Curr Alzheimer Res* 2011;**8**:4−23.

61. Billings LM, Oddo S, Green KN, McGaugh JL, LaFerla FM. Intraneuronal Aβ causes the onset of early Alzheimer's disease-related cognitive deficits in transgenic mice. *Neuron* 2005;**45**:675−88.

62. Oddo S, Caccamo A, Shepherd JD, Murphy MP, Golde TE, Kayed R, et al. Triple-transgenic model of Alzheimer's disease with plaques and tangles: intracellular Aβ and synaptic dysfunction. *Neuron* 2003;**39**:409−21.

63. Casas C, Sergeant N, Itier JM, Blanchard V, Wirths O, van der Kolk N, et al. Massive CA1/2 neuronal loss with intraneuronal and N-terminal truncated Aβ42 accumulation in a novel Alzheimer transgenic model. *Am J Pathol* 2004;**165**:1289−300.

64. Shie F-S, LeBoeur RC, Jin L-W. Early intraneuronal Aβ deposition in the hippocampus of APP transgenic mice. *NeuroReport* 2003;**14**:123−9.

65. D'Andrea MR, Nagele RG. Morphologically distinct types of amyloid plaques point the way to a better understanding of Alzheimer's disease pathogenesis. *Biotech Histochem* 2010;**85**(2):133−47.

66. D'Andrea MR, Nagele RG. MAP-2 immunolabeling can distinguish diffuse from dense-core amyloid plaques in brains with Alzheimer's disease. *Biotech Histochem* 2002;**77**(2):95−103.

67. D'Andrea MD, Cole GM, Ard MD. The microglial phagocytic role with specific plaque types in the Alzheimer disease brain. *Neurobiol Aging* 2004;**25**:675−83.

68. Knobloch M, Konietzko U, Krebs DC, Nitsch RM. Intracellular Aβ and cognitive deficits precede-amyloid deposition in transgenic arcAβ mice. *Neurobiol Aging* 2007;**28**:1297–306.
69. Parihar MS, Brewer CJ. Review. Amyloid beta as a modulator of synaptic plasticity. *J Alzheimers Dis* 2010;**22**(3):741–63.
70. Perez M, Cuadros R, Benıtez MJ, Jimnez JS. Interaction of Alzheimer's disease amyloid βpeptide fragment 25–35 with tau protein, and with a tau peptide containing the microtubule binding domain. *J Alzheimers Dis* 2004;**6**:461–7.
71. Stancu I-C, Vasconcelos B, Terwel D, Dewachter I. Models of β-amyloid induced Tau-pathology: the long and "folded" road to understand the mechanism. *Mol Neurodegener* 2014;**9**:51.
72. Gotz J, Chen F, Van DJ, Nitsch RM. Formation of neurofibrillary tangles in P301l tau transgenic mice induced by Aβ42 fibrils. *Science* 2001;**293**(5534):1491–5.
73. Bolmont T, Clavaguera F, Meyer-Luehmann M, Herzig MC, Radde R, Staufenbiel M, et al. Induction of tau pathology by intracerebral infusion of amyloid-β-containing brain extract and by amyloid-β deposition in APP x Tau transgenic mice. *Am J Pathol* 2007;**171**(6):2012–20.
74. Goedert M. Review. Oskar Fisher and the study of dementia. *Brain* 2009;**132**:1102–11.

Intraneuronal Amyloid and Cognitive Impairment

A myloid deposition is the basis of the amyloid cascade hypothesis, and removing the extracellular Aβ plaques will halt the progression of cognitive decline in AD. In spite of heeding earlier data reporting a weak and inconsistent relationship between cortical senile plaque count and cognitive decline, the clinical trials were launched.[1,2] These trials have unfortunately failed to halt the progress of the cognitive decline in the AD patients thereby challenging its supporters to revise the hypothesis, while leaving others to doubt the significance of amyloid in the pathogenesis of the disease. Perhaps, it is the species of Aβ that is critical for toxicity, leaving the once-toxic Aβ plaques to be not so toxic.[3]

In spite of the removal of amyloid plaques in patients with AD, perhaps the failure to cure the disease can be explained by the significant presence of amyloid plaque loads detected in cognitively normal individuals using live molecular imaging techniques that were verified in the autopsied brain tissues.[4–6] In transgenic mouse AD models, amyloid deposition in senile plaques appeared after cognitive defects.[7–10] In the triple-transgenic rats, significant cognitive impairment, as assessed by the Morris water maze task, occurred at 7 months of age, which was 2 months before the deposition of mature plaques.[11,12]

The numbers of plaques present in the brain did not correlate with the severity of dementia, but amyloid was not out of the picture yet. Recently, a clinical correlation between elevated levels of total Aβ peptide in the brain and cognitive decline was reported.[13] A variety of studies have indicated that elevation and accumulation of Aβ levels result in cognitive dysfunction, including memory deficits.[14]

A revised amyloid model proposed that soluble oligomeric Aβ42 and not plaque-associated Aβ correlated better with cognitive dysfunction in AD.[13,15] Also, Aβ oligomers can preferentially form within the neuronal processes and synapses rather than extracellularly, and that leads to synaptic alterations.[16–19]

INTRANEURONAL Aβ TO COGNITION

Intracellular Levels

Intraneuronal punctate deposits of Aβ occurred concomitantly with robust cognitive impairments at the age of 6 months in the transgenic (Swedish and Arctic mutations) AD mice before the detection of Aβ plaque formation and CAA.[20] Extensive studies in the triple-transgenic mice showed that intraneuronal Aβ is an important pathological trigger for the onset of cognitive deficits rather than the extracellular deposition of Aβ.[21] In another study using the same mice, cognitive impairment as defined as a deficit in long-term retention, correlated with accumulation of intraneuronal Aβ in the hippocampus and amygdala well before the detection of plaques or tangles at 4 months of age.[21] Clearance of the intraneuronal Aβ by immunotherapy rescued the cognitive impairments that returned as the intraneuronal Aβ accumulated. Significant neuronal loss (~30% CA1 region/18% hippocampus atrophy) in the 1-year-old APP/PS1KI mice correlated with intraneuronal accumulation of various Aβ peptides before extracellular plaque pathology that exhibited severe and early learning deficits at 6 months of age.[22–24]

In a water maze learning task, male transgenic rats with the Swedish and Indiana mutations displayed a mild spatial learning deficit relative to control littermates that correlated with intraneuronal accumulation of Aβ peptide in the absence of plaques.[25,26] Intraneuronal can also oligo-merize forming cytotoxic aggregated that can induce cognitive impair-ments (see Chapter 5).[27,28]

Extracellular Aβ can bind quite effectively to the α7 receptor with very high affinity, which normally regulate calcium homeostasis and acetylcholine release, two key events in cognition and memory. Excessive binding suggests that the α7 receptor mediates at least some of the toxic effects of intraneuronal Aβ that can lead to cognitive impairment.[29]

Correlation of Aβ-Induced Synaptic Changes With Cognition

Synapse alterations and loss are the strongest anatomical neuropathological features that correlate with the degree of clinical impairment in AD.[30,31] Cognitive decline correlates better with synapse loss or tau pathology than with Aβ plaque burden.[2,30] Alterations in synaptic plasticity, mitochondrial function, and neurotransmission by Aβ can affect activity-dependent signaling and gene expression resulting in the disintegration of neural networks and ultimately, the failure of neural functions.[32]

However, extracellular soluble Aβ at nM concentrations can account for synaptic damage in vitro.[33] In the triple-transgenic AD mice, deficits in synaptic transmission and LTP leading to cognitive deficits correlated with early intraneuronal Aβ accumulation.[13,34] Continuous overproduction of Aβ at dendrites or axons can reduce the number and plasticity of synapses.[35,36] Electron microscopic studies demonstrate that oligomeric Aβ is localized within the synaptic compartment suggesting that oligomeric Aβ may interact directly at the synapse to cause dysfunction and spine collapse.[16,37]

In a diet-induced model of hypercholesterolemia in the transgenic mice, cognitive dysfunction was accompanied with hippocampal synaptic loss and intraneuronal Aβ accumulation.[38]

SUMMARY

Although discussed in previous chapters, one of the pathological consequences of intraneuronal Aβ that may begin with synaptic decline is cognitive impairment. There is no correlation of extracellular Aβ plaques to cognitive changes.

References

1. Crystal H, Dickson D, Fuld P, et al. Clinico-pathologic studies in dementia: nondemented subjects with pathologically confirmed Alzheimer's disease. *Neurology* 1988;**38**:1682—7.
2. Arriagada PV, Growdon JH, Hedley-Whyte ET, Hyman BT. Neurofibrillary tangles but not senile plaques parallel duration and severity of Alzheimer's disease. *Neurology* 1992;**42**:631—9.

3. Pimplikar SW. Review. Reassessing the amyloid cascade hypothesis of Alzheimer's disease. *Int J Biochem Cell Biol* 2009;**41**:1261–8.
4. Wischik CM, Wischik DJ, Storey JMD, et al. Review. Rationale for tau aggregation inhibitor therapy in Alzheimer's disease and other tauopathies. *RSC Drug Discov [Online]* 2010;1.
5. Nordberg A. Review. Amyloid plaque imaging in vivo: current achievement and future prospects. *Eur J Nucl Med Mol Imaging* 2008;**35**(S1):S46–50.
6. Villemagne VL, Fodero-Tavoletti MT, Pike KE, et al. The ART of loss: Aβ imaging in the evaluation of Alzheimer's disease and other dementias. *Mol Neurobiol* 2008;**38**:1–15.
7. Hsia AY, Masliah E, McConlogue L, Yu G-Q, Tatsuno G, Hu K, et al. Plaque-independent disruption of neural circuits in Alzheimer's disease mouse models. *PNAS* 1999;**96**:3228–33.
8. Holcomb L, Gordon MN, McGowan E, Yu X, Benkovic S, Jantzen P, et al. Accelerated Alzheimer-type phenotype in transgenic mice carrying both mutant amyloid precursor protein and presenilin 1 transgenes. *Nat Med* 1998;**4**:97–100.
9. Moechars D, Dewachter I, Lorent K, Reverse D, Baekelandt V, Naidu A, et al. Early phenotypic changes in transgenic mice that overexpress different mutants of amyloid precursor protein in brain. *J Biol Chem* 1999;**274**:6483–92.
10. Chui DH, Tanahashi H, Ozawa K, Ikeda S, Checler F, Ueda O, et al. Transgenic mice with Alzheimer presenilin 1 mutations show accelerated neurodegeneration without amyloid plaque formation. *Nat Med* 1999;**5**:560–4.
11. Flood DG, Lin YG, Lang DM, Trusko SP, Hirsch JD, Savage MJ, et al. A transgenic rat model of Alzheimer's disease with extracellular Aβ deposition. *Neurobiol Aging* 2007;**30**:1078–90.
12. Liu L, Orozco IJ, Planel E, Wen Y, Bretteville A, Krishnamurthy P, et al. A transgenic rat that develops Alzheimer's disease-like amyloid pathology, deficits in synaptic plasticity and cognitive impairment. *Neurobiol Dis* 2008;**31**:46–57.
13. Näslund J, Haroutunian V, Mohs R, et al. Correlation between elevated levels of amyloid β-peptide in the brain and cognitive decline. *JAMA* 2000;**283**:1571–7.
14. Haass C, Selkoe DJ. Review. Soluble protein oligomers in neurodegeneration: Lessons from the Alzheimer's amyloid β-peptide. *Nat Rev Mol Cell Biol* 2007;**8**:101–12.
15. McLean CA, Cherny RA, Fraser FW, Fuller SJ, Smith MJ, Beyreuther K, et al. Soluble pool of Aβ amyloid as a determinant of severity of neurodegeneration in Alzheimer's disease. *Ann Neurol* 1999;**46**:860–6.
16. Takahashi RH, Almeida CG, Kearney PF, Yu F, Lin MT, Milner TA, et al. Oligomerization of Alzheimer's β-amyloid within processes and synapses of cultured neurons and brain. *J Neurosci* 2004;**24**:3592–9.
17. Takahashi RH, Milner TA, Li F, Nam EE, Edgar MA, Yamaguchi H, et al. Intraneuronal Alzheimer aβ42 accumulates in multivesicular bodies and is associated with synaptic pathology. *Am J Pathol* 2002;**161**:1869–79.
18. Lacor PN, Buniel MC, Chang L, Fernandez SJ, Gong Y, Viola KL, et al. Synaptic targeting by Alzheimer's-related amyloid beta oligomers. *J Neurosci* 2004;**24**:10191–200.
19. Lacor PN, Buniel MC, Furlow PW, Sanz Clemente A, Velasco PT, Wood M, et al. Aβ oligomer-induced aberrations in synapse composition, shape, and density provide a molecular basis for loss of connectivity in Alzheimer's disease. *J Neurosci* 2007;**27**:796–807.
20. Knobloch M, Konietzko U, Krebs DC, Nitsch RM. Intracellular Aβ and cognitive deficits precede—amyloid deposition in transgenic arcAβ mice. *Neurobiol Aging* 2007;**28**:1297–306.
21. Billings LM, Oddo S, Green KN, McGaugh JL, LaFerla FM. Intraneuronal Aβ causes the onset of early Alzheimer's disease-related cognitive deficits in transgenic mice. *Neuron* 2005;**45**:675–88.

22. Casas C, Sergeant N, Itier JM, Blanchard V, Wirths O, et al. Massive CA1/2 neuronal loss with intraneuronal and N-terminal truncated Aβ42 accumulation in a novel Alzheimer transgenic model. *Am J Pathol* 2004;**165**:1289–300.

23. Christensen DZ, Kraus SL, Flohr A, Cotel MC, Wirths O, Bayer TA. Transient intraneuronal Aβ rather than extracellular plaque pathology correlates with neuron loss in the frontal cortex of APP/PS1KI mice. *Acta Neuropathol* 2008;**116**:647–55.

24. Wirths O, Breyhan H, Schafer S, Roth C, Bayer TA. Deficits in working memory and motor performance in the APP/PS1ki mouse model for Alzheimer's disease. *Neurobiol Aging* 2008;**29**:891–901.

25. Leon WC, Canneva F, Partridge V, Allard S, Ferretti MT, DeWilde A, et al. A novel transgenic rat model with a full Alzheimer's-like amyloid pathology displays pre-plaque intracellular amyloid-β-associated cognitive impairment. *J Alzheimers Dis* 2010;**20**:113–16.

26. Echeverria V, Ducatenzeiler A, Dowd E, Jänne J, Grant SM, Szyf M, et al. Altered mitogen-activated protein kinase signaling, tau hyperphosphorylation and mild spatial learning dysfunction in transgenic rats expressing the β-amyloid peptide intracellularly in hippocampal and cortical neurons. *Neuroscience* 2004;**129**(3):583–92.

27. Cleary JP, Walsh DM, Hofmeister JJ, Shankar GM, Kuskowski MA, Selkoe DJ, et al. Natural oligomers of the amyloid-β protein specifically disrupt cognitive function. *Nat Neurosci* 2005;**8**:79–84.

28. Lesne S, Koh MT, Kotilinek L, Kayed R, Glabe CG, Yang A, et al. A specific amyloid-β protein assembly in the brain impairs memory. *Nature* 2006;**440**:352–7.

29. Mohamed A, de Chaves EP. Review. Aβ internalization by neurons and glia. *Int J Alzheimers Dis* 2011;**127984**:17 pp.

30. Terry RD, Masliah E, Salmon DP, Butters N, DeTeresa R, Hill R, et al. Physical basis of cognitive alterations in Alzheimer's disease: synapse loss is the major correlate of cognitive impairment. *Ann Neurol* 1991;**30**:572–80.

31. Coleman PD, Yao PJ. Synaptic slaughter in Alzheimer's disease. *Neurobiol Aging* 2003;**24**:1023–7.

32. Palop JJ, Chin J, Mucke L. A network dysfunction perspective on neurodegenerative diseases. *Nature* 2006;**443**:768–73.

33. Walsh DM, Selkoe DJ. Review. Aβ oligomers—a decade of discovery. *J Neurochem* 2007;**101**:1172–84.

34. Nilsberth C, Westlind-Danielsson A, Eckman CB, Condron MM, Axelman K, Forsell C, et al. The 'Arctic' APP mutation (E693G) causes Alzheimer's disease by enhanced Aβ protofibril formation. *Nat Neurosci* 2001;**4**:887–93.

35. Shankar GM, Bloodgood BL, Townsend M, Walsh DM, Selkoe DJ, Sabatini BL. Natural oligomers of the Alzheimer amyloid-β protein induce reversible synapse loss by modulating an NMDA-type glutamate receptor-dependent signaling pathway. *J Neurosci* 2007;**27**:2866–75.

36. Wei W, Nguyen LN, Kessels HW, Hagiwara H, Sisodia S, Malinow R. Amyloid beta from axons and dendrites reduces local spine number and plasticity. *Nat Neurosci* 2010;**13**:190–6.

37. Koffie RM, Meyer-Luehmann M, Hashimoto T, Adams KW, Mielke ML, Garcia-Alloza M, et al. Oligomeric amyloid beta associates with postsynaptic densities and correlates with excitatory synapse loss near senile plaques. *PNAS* 2009;**106**:4012–17.

38. Umeda T, Tomiyama T, Kitajima E, Idomoto T, Nomura S, Lambert MP, et al. Hypercholesterolemia accelerates intraneuronal accumulation of Aβ oligomers resulting in memory impairment in Alzheimer's disease model mice. *Life Sci* 2012;**91**:1169–76.

CHAPTER

7

Intraneuronal Amyloid and Inflammation

A lzheimer's disease brains are plagued with neuroinflammation, which has also been observed in brains of the transgenic mouse models of AD. Actually about 100 years ago, Oskar Fischer noted that extracellular deposition was the critical step in plaque formation that provokes neuroinflammation, which is similar to present-day amyloid deposition model of AD that Aβ secreted from neurons is toxic, ultimately leading to a cascade of neurotoxic events in AD (see Chapter 4).[1–3] Neuroinflammation in the brain is typically defined as gliosis, a reparative response where reactive microglia, the resident phagocyte, and activated astrocytes work together in tandem to clear debris and/or foreign material leading to the formation of the glial scar.[4]

The accepted view is that the amyloid plaques are formed through chronic inflammatory responses induced by fibrillar deposits of Aβ, as

Intracellular Consequences of Amyloid in Alzheimer's Disease.
DOI: http://dx.doi.org/10.1016/B978-0-12-804256-4.00007-3

reactive microglia are typically in association with amyloid deposits. Since the extracellular deposition of toxic Aβ plaques occurs well after cognitive impairment, the consequences of neuroinflammation would not be expected to significantly impact cognitive decline (see Chapter 6).[2] But then how to explain the beneficial effects of nonsteroidal anti-inflammation drugs (NSAIDs) in reducing the risk of developing AD when cognitive decline occurs prior to neuroinflammation?

EFFICACY OF NSAIDs

The evidence from numerous epidemiological studies shows that AD can be prevented by blocking inflammatory reactions with NSAIDs that develop during the course of the disease.[5–9] NSAIDs are a category of medications that include the salicylate, propionic acid, acetic acid, fenamate, oxicam, and the cyclooxygenase (COX)-2 inhibitor classes.[3] Over 20 epidemiological studies determined that anti-inflammatory drugs like indomethacin and ibuprofen reduce the risk of AD.[10–12] Similarly, a decreased risk of AD was observed in patients with rheumatoid arthritis and osteoarthritis who were treated with NSAIDs for long periods of time.[9] Although studies appear to show that NSAIDs can prevent the risk of AD, the results of clinical trials with anti-inflammatory drugs in AD patients were negative especially for the COX-2 inhibitors.[13–16]

Perhaps the inconsistent findings lie in the inaccuracy of the pathological sequences leading to AD based on the amyloid deposition model. In a number of preclinical transgenic mouse models of AD, the beneficial effects of NSAIDs demonstrated that even a short-term treatment of NSAIDs can significantly lower soluble levels of Aβ42, delay AD symptoms, decrease Aβ deposition, and improve behavioral tasks.[4–8] These beneficial results can be explained through the analyses of autopsied brain tissues from nondemented patients. Long-term NSAID usage effectively and significantly reduced the numbers of microglia (~threefold) without affecting the number of diffuse or neuritic (dense-core) plaques in nondemented patients as compared with similar patients with no history of arthritis or other condition that might promote regular NSAIDs usage.[17] Other than the finding of plaques in normal patients, these data show that NSAIDs effectively suppress microglial activation. Interestingly, the NSAIDs did not affect plaque formation. The latter observation appears to nullify the widely accepted belief that inflammatory cells contribute to plaque formation, and the hypothesis that diffuse plaques mature into the dense-core/ neuritic plaques over time due to the processes of inflammation. If the inflammatory cells (microglia and astrocytes) help mature diffuse amyloid deposits into neuritic, dense-core plaques, then one would expect fewer plaques in the patients who were treated with long-term doses of NSAIDs, which was not reported.

The limited efficacy of NSAIDs to treat patients with AD may also lie in the inability to suppress the primary pathological event, which is the lysing of the Aβ-overburdened neurons, while benefits of NSAID therapies could help to offset subsequent secondary neuronal death due to the secreted factors from the reactive microglia and activated astrocytes.[18-20] Even as far back as 1898, it was believed that plaques corresponded to a modified type of glial cell, mostly because of the presence of fibrous material. Then it was concluded that glial cell proliferation was a secondary, not primary event to nerve cell degeneration.[1] These findings indicate that the anti-inflammatory agents can be helpful in the prevention, and perhaps averting further cognitive decline, but not in the treatment of AD because the neurons that elicited the inflammatory response are already dead.[19-21] Clinical trials are currently testing promising new NSAIDs and combination of NSAIDs for the treatment of AD.[22]

INTRANEURONAL Aβ's ROLE IN GLIOSIS

In contrast to the dogmatic amyloid deposition-based hypothesis, there are particular plaque types with unique origins of formation suggesting that not all extracellular Aβ is toxic in activating an immune response. The dense-core Aβ42 senile plaques (thioflavin S-positive) are associated with reactive microglia and activated astrocytes (gliosis) while the diffuse Aβ42 plaques (thioflavin S-negative) do not promote an inflammatory reaction.[19,20,23,24] MAP2-positive neuronal processes were obliterated in the wake of the dense-core plaques (Fig. 7.1) that were unaffected in areas of the diffuse plaques.[25] The destruction of brain elements in the area defined by the size of the dense-core plaques can only be explained by an acute event (eg, lysed neuron) because chronic extracellular depositions of Aβ42 rarely decorate morphologically healthy or even degenerating neurons. Neuronal debris such as DNA, cathepsin D, and ubiquitin are not detected in areas of the diffuse-type plaques that are not associated with inflammatory cells (see Chapter 9).[20,26]

The link between intraneuronal Aβ and gliosis is time-dependent: the overaccumulation of intraneuronal Aβ leads to cell death, which then triggers gliosis.[20] It is an "Inside-Out" phenomenon: Aβ-overburdened neurons lyse or burst from the "inside" leaching "out" neuronal

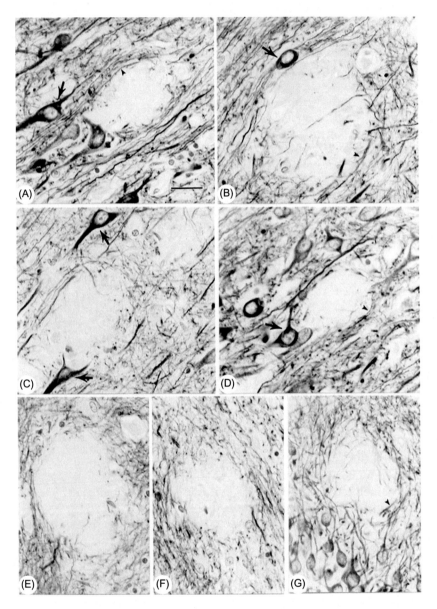

FIGURE 7.1   Areas without MAP2 immunolabeling in the Alzheimer's disease entorhinal cortex (A–D) and hippocampus (E–G) clearly show the presence of some pathological issue, determined to be Aβ42-positive dense-core plaques. Note the unexplained presence of intense MAP2 immunolabeling in neurons in the vicinity of local, MAP2-deficient regions (*arrows*). Arrowheads indicate MAP-2-positive neuronal processes that appear to bend around the margins of these areas. Scale bar, 25 μm. *Aβ42*, amyloid-β42; *MAP2*, microtubule-associated protein 2. *Source: Used with permission from* Biotech Histochem *2002;**77**(2):95–103.*

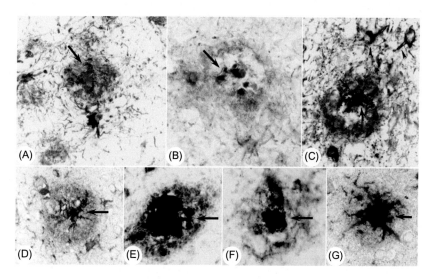

FIGURE 7.2   Representative images of triple immunohistochemical labeling in AD cortical tissues showing the presence of purple-stained, HLA-DR-positive, reactive microglia (*arrows*) in the middle of these Aβ42-positive (*red*) amyloid plaques. Black-labeled (A, C) or brown-labeled (B, D, F) GFAP-activated astrocytes are observed. *Aβ42*, amyloid-β42; *HLA-DR*, human leukocyte antigen-DR; *GFAP*, glial fibrillary acidic protein. *Source: Used with permission from* Biotech Histochem 2001;76(2):97−106; Elsevier Press 2014.

debris including DNA, ATP, active lysosomes, etc., that trigger gliosis (see Chapter 9).[18,20,26,27]

Location of the inflammatory cells in the dense-core, senile plaques may help explain their presence. The reported function of microglia is to remove or clear extracellular deposits of toxic Aβ through receptor-mediated phagocytosis.[28,29] Phagocytic microglia also participate in the removal of degenerating neurons and synapses as well as Aβ deposits.[19,30] One would expect reactive microglia to surround the periphery of dense-core plaques as the microglia would begin removing toxic Aβ from the perimeter first.[23] But this was not observed using triple immunolabeling methods to simultaneously detect Aβ42, astrocytes, and microglia on AD brain tissues (Fig. 7.2).[26]

The reactive microglia were detected mostly in the centers of these dense-core plaques suggesting that they are attracted by material in the center such as DNA, rather than the plaque amyloid. Additional evidence shows that microglia are co-localized with the neuronal nuclear protein NeuN (Fig. 7.3).[24,26,31] Extracellular DNA and ATP released as the neuron lyses (or burst) are chemotaxis activators of microglia through the G protein-coupled P2Y and/or scavenger receptor A (CD36) microglial receptors.[32,33]

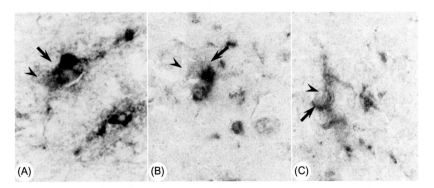

FIGURE 7.3   Representative images of the Alzheimer's disease cortical tissues processed for a double immunohistochemical assay show reactive HLA-DR-positive microglia in purple (*arrows*) in close association with red-labeled NeuN-positive neurons (*arrowheads*). HLA-DR, human leukocyte antigen-DR; *NeuN*, neuronal-specific neuronal protein. *Source: Used with permission from* Biotech Histochem *2010;85(5):133−47.*

This possibility also explains why microglia are not associated with DNA-negative and diffuse plaques that are proposed to originate from leaky, dysfunctional blood vessels (Fig. 5.4). The lack of microglia with diffuse plaques provides additional evidence invalidating the original teachings of the amyloid deposition-based hypothesis that diffuse plaques evolve over time with the contributions of gliosis to mature into dense-core plaques. Additional evidence from transgenic mice and from nondemented patients with amyloid plaques suggests that fibrillar Aβ alone may not be sufficient to initiate brain inflammation.[4]

The most likely initiator of neuroinflammation is neuronal death although other factors such as an age-related increase in steady-state levels of brain inflammation, or severe or subtle traumatic brain injury, resulting from blunt trauma or cardiovascular disease, might initiate inflammation that is exacerbated by the presence of amyloid plaques.[34]

INFLAMMATORY CELLS EXACERBATE AD

The subsequent consequences of the inflammatory cells can further worsen the extent of neuropathology in AD. As in the case of the peripheral system, the inflammatory response in the brain can also do as much harm as good. The onset of AD coincides with the detection of inflammatory markers around amyloid plaques and dystrophic neurites. Each of these CNS-inflammatory cells (microglia and astrocytes) secretes a number of factors that can unfortunately harm locally functioning neurons based on sets of reports that support the idea that altered patterns in the glia—neuron interactions constitute early molecular events leading to neurodegeneration in AD.[35] A direct correlation has been established between the Aβ-induced neurodegeneration and cytokine production, and its subsequent release. Neuroinflammation is responsible for an abnormal secretion of proinflammatory cytokines, chemokines, and complement activation products from the resident CNS cells that trigger signaling pathways, and play a relevant role in the pathogenesis of the inflammatory process occurring during the development of the pathology because of their chemotactic activity on brain phagocytes.[35–38]

In vitro studies show that Aβ can activate cultured astrocytes and microglia cells, presumably through Aβ interactions with cell-surface receptors.[5] The inflammatory events may also result from Aβ-mediated activation of the complement system in the CNS. However, this was not observed for the diffuse-type amyloid plaques in AD brain tissues.[19,20,24]

Reactive Microglia

The microglia are the resident macrophages in the brain, and are the first line of defense that constantly scavenge the brain for infectious agents, damaged or dead cells and in the case of AD, are attracted to amyloid plaques, specifically the dense-core, senile plaques. The presence of amyloid plaques with reactive microglia and no activated astrocytes implies newly formed amyloid plaques (recent pyramidal neuron deaths).[26] Neuronal death then can trigger a cascade of inflammatory responses beginning with microglial migration, and subsequent microglial phagocytosis, which then secrete factors that activate local astrocytes in the processes of gliosis.[26] Continued activation of microglia in these lesions elicits a persistent inflammatory response.[39] However, heavy accumulation of pathological plaque debris in post-mortem indicates the failure, or at best partial success, of the removal.

While some microglial functions are beneficial, the destructive effects of the production of toxins (such as nitric oxide and superoxide), and proinflammatory cytokines by reactive microglia apparently overcome the protective functions in the chronic stage of neuroinflammation.[40,41] In vitro studies show that microglia can elicit protective and toxic effects in response to Aβ, which may depend on the state of activation of the microglia.[42,43] In addition to phagocytosis (particularly of dead cells or neuronal blebs), microglial can induce neuronal C1q synthesis perhaps as an early response to injury to facilitate clearance of damaged or degenerating cells, while modulating inflammation and perhaps facilitating repair.[44,45] C1q was co-localized with immunoglobulin-positive, degenerating neurons in AD tissues providing evidence that some of the neurons in the AD brain are dying through the classical antibody-dependent complement pathway.[46] Activated microglia were also spatially more associated with these immunoglobulin-positive neurons than with immunoglobulin-negative neurons suggesting additional neuronal death through an autoimmune mechanism in AD.[20,47,48]

Microglial activation involves the induction of CD45, CD40, CR3, and IL-8; all of which correlate with intraneuronal accumulation of Aβ indicating contribution of factors released by neurons upon Aβ exposure.[49] Monocyte chemotactic protein-1 (MCP-1) is essential for monocyte recruitment in inflammatory models in vivo. MCP-1 −/− mice are unable to recruit monocytes.[50] In these mice, only mature senile plaques were detected suggesting that MCP-1-related inflammatory events induced by reactive microglia contribute to the maturation of senile plaques.[51] Also, MCP-1 levels are higher in CSF of patients with AD than in controls suggesting a role of phagocytic cells within the brain during the development of dementia.[52,53]

Expression analysis of RAGE in nondemented and AD brains indicated that increases in RAGE protein and percentage of RAGE-expressing microglia paralleled the severity of the disease.[54] The adverse consequences of RAGE interaction with Aβ include perturbation of neuronal properties and functions, amplification of glial inflammatory responses, elevation of oxidative stress and amyloidosis, increased Aβ influx at the BBB and vascular dysfunction, and induction of autoantibodies.

A number of in vitro studies have further demonstrated that microglia are responsible for initiating and sustaining inflammatory cascades that involve astrocyte activation contributing to neurodegeneration in AD.[54] Exposing human microglia to preaggregated Aβ42 (0.1 to 10 millimolar) resulted in secretion of inflammatory cytokines, chemokines, and nitric oxide.[55] Of these, the macrophage colony stimulating factor, which is important for neuronal survival, microglial growth and activation, is elevated fivefold in the CSF of AD patients, and is upregulated in activated microglia as well as in injured neurons, which signal microglia recruitment.[55–57]

Cultured microglia cells also express the α7 receptor that could also explain the presence of intracellular Aβ independent of the processes of phagocytosis.[58]

Activated Astrocytes

Some of the vital functions of the astrocytes in the brain are structural as well as biochemical. In the synapses, the astrocytes help regulate the transmission of electrical impulses from neuron to neuron, and in the BBB, they help provide the barrier function with the endothelial cells. The astrocytes regulate ion concentrations in the extracellular space, help provide nutrients such as glycogen and lactate to neurons, regulate blood flow to the neurons, and interact with microglia in the processes of neuroinflammation.[4]

In the processes of gliosis, the astrocytes become activated largely by proliferation, and hypertrophy of their cell processes and bodies with increased expression of the cytoskeletal intermediate filament, glial fibrillary acidic protein.[59] The activated astrocytes extend their processes in an attempt to encapsulate the damaged area forming the glial scar, perhaps in an effort to quarantine the area. The local consequences of gliosis can impede neuronal dendritic growth, and interfere with eventual reconnection of functional neuronal circuitry by inhibiting axonal regeneration, preventing remyelination, or promotion of abnormal neuronal connections with increased seizure foci.[60] One of the secreted factors from the activated astrocytes is vascular endothelial growth factor that can also break down the BBB (see Chapter 2).[59]

Although the cytokines secreted by activated microglia will then activate astrocytes, Aβ can induce functional and morphological changes of activated astrocytes without being toxic to them.[61–64] Aβ promotes phosphorylation and nuclear translocation of the extracellular signal-regulated kinases 1 and 2 that regulate gene expression in cultured rat cortical astrocytes.[65]

Astrocytic pathology is also correlated with impaired cerebral metabolism in transgenic mice, and that astrocyte alterations occur already at premature stages of pathology suggesting that astrocyte dysfunction can contribute to early behavioral and cognitive impairments seen in these mice.[66]

Like some neurons, astrocytes also express the α7 receptor, which may also contribute to their demise as an astrocytic plaque through the toxic accumulation of Aβ (see Chapter 2).[67] Significant increases in the total numbers of astrocytes and of astrocytes expressing the α7 receptors were observed in the hippocampus and temporal cortex of both transgenic APPSwe mice and sporadic AD brains as compared to normal brain tissues.[68] The increase in the level of expression of α7 receptors on astrocytes positively correlated with the extent of neuropathological alternations, especially with the number of neuritic plaques in the AD brain.

Astrocytes throughout the entorhinal cortex of AD patients gradually accumulate Aβ42-positive material, and the amount of this material correlates positively with the extent of local AD pathology. It is not clear if the intracellular Aβ in the astrocytes is mediated through the α7receptor and/or through phagocytosis. Like that reported for Aβ-overburdened neurons, some astrocytes containing Aβ42-positive deposits can also undergo lysis, resulting in the formation of astrocyte-derived amyloid plaques in affected brain regions.[67]

SUMMARY

The numbers of neurotoxic consequences from the activated inflammatory cells in the AD brain are well beyond the scope of this book, but certainly make the point that in an effort to do well, they cause more harm. A proposed model representing some of these events is presented in Fig. 7.4 and shows the vicious cycle of events leading to additional cell death in the AD brain. Once the neurodegeneration cascade is initiated, microglial and astrocytes may play major roles directly and indirectly promoting self-sustaining neurodegeneration cycles.[19,36,39,69]

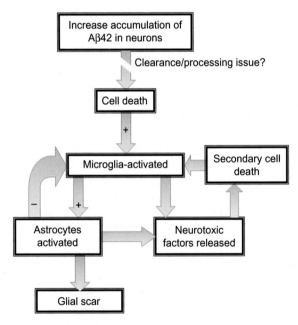

FIGURE 7.4   The Inflammatory Cascade model. A proposed pathway of inflammation in the AD brain showing inhibitory (−) and stimulatory (+) pathways of primary and secondary cell death consequences. Based on substantial evidence, it was proposed that one of the primary activators of microglial activations originates from dying or lysed cells. Subsequent microglial activation can trigger astrocyte activation (+) (although astrocytic activation may occur independent of microglial activation), which in turn can inhibit microglial activation (−). However, both reactive microglia and astrocytes secrete factors that are toxic to neurons, thereby contributing to a pathological cascade. Aβ42, amyloid-β42. *Source: Used with permission from* Neurobiol Aging 2004;25:675–83.

These inflammatory cells can also secrete Aβ40 and Aβ42 as well as cytokines including tumor necrosis factor-α and interferon-γ increase levels of endogenous BACE1, APP, and Aβ, that stimulate additional amyloidogenic APP processing in astrocytes.[3,70,71] Oligomeric and fibrillar Aβ42 also increases levels of astrocytic BACE1, APP, and β-secretase processing.[71] Given that astrocytes greatly outnumber neurons, activated astrocytes may represent significant sources of Aβ during neuroinflammation in AD, and due to these high levels, it is believed that they can contribute significantly to the formation of amyloid plaques in AD.[72,73]

Proinflammatory molecules secreted from activated astrocytes can subsequently upregulate the expression of secretases in neurons to further enhance Aβ production, and can activate microglia to produce additional inflammatory factors.[74–76]

Aβ-induced inflammation can also contribute to Aβ-induced tau pathology. The cascade of events initiated by the reactive microglia may indirectly result in hyperphosphorylation and self-aggregation of tau protein into neurotoxic oligomeric species.[77] Consistent with this result is the fact that blocking of interleukin (IL)-1 signaling with an IL-1-R antibody attenuated tau pathology in triple-transgenic mice, while increasing IL-1β exacerbated tau pathology.[78–79]

When considering the many toxic factors that can directly and/or indirectly kill neurons, it is no wonder why no one has yet to clearly define the first detrimental circumstance that triggers the cascade of pathological events leading to AD.

References

1. Goedert M. Review. Oskar Fisher and the study of dementia. *Brain* 2009;**132**:1102–11.
2. Eikelenboom P, Veerhuis R, Scheper W, Rozemuller AJ, van Gool WA, Hoozemans JJ. Review. The significance of neuroinflammation in understanding Alzheimer's disease. *J Neural Trans* 2006;**113**(11):168595.
3. Tuppo EE, Arias HR. Review. The role of inflammation in Alzheimer's disease. *Int J Biochem Cell Bio* 2005;**37**:289–305.
4. https://en.wikipedia.org/wiki/Gliosis [accessed 10.10.15].
5. Emmerling MR, Watson MD, Raby CA, Spiegel K. Review. The role of complement in Alzheimer's disease pathology. *Biochimica et Biophysica Acta* 2000;**1502**:158–71.
6. Sastre M, Klockgether T, Heneka MT. Review. Contribution processes to Alzheimer's disease: molecular mechanisms. *Int J Neurosci* 2006;**24**:167–76.
7. Weggen S, Eriksen JL, Das P, Sagi SA, Wang R, Pietrzik CU, et al. A subset of NSAIDs lower amyloidogenic Aβ42 independently of cyclooxygenase activity. *Nature* 2001;**414**:212–16.
8. Eriksen JL, Sagi SA, Smith TE, Weggen S, Das P, McLendon DC, et al. NSAIDs and enantiomers of flurbiprofen target γ-secretase and lower Aβ42 in vivo. *J Clin Invest* 2003;**112**:440–9.
9. Lim GP, Yang F, Chu T, Chen P, Beech W, Teter B, et al. Ibuprofen suppresses plaque pathology and inflammation in a mouse model for Alzheimer's disease. *J Neurosci* 2000;**20**:5709–14.

10. McGeer PL, Schulzer M, McGeer EG. Review. Arthritis and anti-inflammatory agents as possible protective factors for Alzheimer's disease: a review of 17 epidemiologic studies. *Neurology* 1996;**47**:425−32.
11. McGeer EG, McGeer PL. Review. The importance of inflammatory mechanisms in Alzheimer disease. *Exp Gerontol* 1998;**33**:371−8.
12. McGeer PL, McGeer EG. Review. Inflammation of the brain in Alzheimer's disease: implications for therapy. *J Leukocyte Biol* 1999;**65**:409−15.
13. Güntert A, Döbeli H, Bohrmann B. High sensitivity analysis of amyloid β peptide composition in amyloid deposits from human and PS2APP mouse brain. *Neuroscience* 2006;**143**(2):46175.
14. Aisen PS. Review. The potential of anti-inflammatory drugs for the treatment of Alzheimer's disease. *Lancet Neurol* 2002;**1**:279−84.
15. Aisen PS, Davis KL, Berg JD, Schafer K, Campbell K, Thomas RG, et al. A randomized controlled trial of prednisone in Alzheimer's disease: Alzheimer's disease cooperative study. *Neurology* 2000;**54**:588−93.
16. Aisen PS, Schafer KA, Grundman M, Pfeiffer E, Sano M, Davis KL, et al. Effects of rofecoxib or naproxen vs placebo on Alzheimer disease progression. *JAMA* 2003;**289**:2819−26.
17. Mackenzie IRA, Munoz DG. Nonsteroidal anti-inflammatory drug use and Alzheimer-type pathology in aging. *Neurology* 1998;**50**(4):986−90.
18. https://www.youtube.com/watch?v = _NTaGjQow1c [accessed 07.10.15].
19. D'Andrea MR, Cole GM, Ard MD. The microglial phagocytic role with specific plaque types in the Alzheimer disease brain. *Neurobiol Aging* 2004;**25**:675−83.
20. D'Andrea MR. *Bursting neurons and fading memories.* New York: Elsevier Press; 2014.
21. Eikelenboom P, van Gool WA. Review. Neuroinflammatory perspectives on the two faces of Alzheimer's disease. *J Neural Trans* 2004;**111**(3):28194.
22. Galimberti D, Scarpini E. Review. Disease-modifying treatments for Alzheimer's disease. *Ther Adv Neurol Disord* 2011;**4**:203−16.
23. Leon WC, Canneva F, Partridge V, Allard S, Ferretti MT, DeWilde A, et al. A novel transgenic rat model with a full Alzheimer's-like amyloid pathology displays pre-plaque intracellular amyloid-β-associated cognitive impairment. *J Alzheimers Dis* 2010;**20**:113−16.
24. D'Andrea MR, Nagele RG. Morphologically distinct types of amyloid plaques point the way to a better understanding of Alzheimer's disease pathogenesis. *Biotech Histochem* 2010;**85**(2):133−47.
25. D'Andrea MR, Nagele RG. MAP-2 immunolabeling can distinguish diffuse from dense-core amyloid plaques in brains with Alzheimer's disease. *Biotech Histochem* 2002;**77**(2):95−103.
26. D'Andrea MR, Reiser PA, Gumula NA, Hertzog BM, Andrade-Gordon R. Application of triple immunohistochemistry to characterize amyloid plaque-associated inflammation in brains with Alzheimer's disease. *Biotech Histochem* 2001;**76**(2):97−106.
27. D'Andrea MR, Lee DHS, Wang H-Y, Nagele RG. Targeting intracellular Aβ42 for Alzheimer's disease drug discovery. *Drug Dev Res* 2002;**56**:194−200.
28. Webster SD, Yang AJ, Margol L, Garzon-Rodriguez W, Glabe CG, Tenner AJ. Complement component C1q modulates the phagocytic behavior of microglia. *Exp Neurol* 2000;**161**:127−38.
29. Rogers J, Strohmeyer R, Kovelowski CJ, Li R. Microglia and inflammatory mechanisms in the clearance of amyloid β peptide. *Glia* 2002;**40**:260−9.
30. Petersen MA, Dailey ME. Diverse microglial motility behaviors during clearance of dead cells in hippocampal slices. *Glia* 2004;**46**:195−206.

31. D'Andrea MR, Nagele RG, Wang H-Y, Peterson PA, Lee DHS. Evidence that neurones accumulating amyloid can undergo lysis to form amyloid plaques in AD. *Histopathology* 2001;**38**:120−34.
32. Erb L, Cao C, Ajit D, Weisman GA. Review. P2Y receptors in Alzheimer's disease. *Biol Cell* 2015;**107**(1):121.
33. Honda S, Sasaki Y, Ohsawa K, Imai Y, Nakamura Y, Inoue K, et al. Extracellular ATP or ADP induce chemotaxis of cultured microglia through Gi/o-coupled P2Y receptors. *J Neurosci* 2001;**21**(6):1975−82.
34. Rozovsky I, Finch CE, Morgan TE. Age-related activation of microglia and astrocytes: in vitro studies show persistent phenotypes of aging, increased proliferation, and resistance to down-regulation. *Neurobiol Aging* 1998;**19**:97−103.
35. Rojo LE, Fernández JA, Maccioni AA, Jimenez JM, Maccioni RB. Review. Neuroinflammation: implications for the pathogenesis and molecular diagnosis of Alzheimer's disease. *Arch Med Res* 2008;**39**(1):116.
36. Liu L, Chan C. Review. The role of inflammasome in Alzheimer's disease. *Ageing Red Rev* 2014;**6**−15.
37. Xia MQ, Hyman BT. Review. Chemokines/chemokine receptors in the central nervous system and Alzheimer's disease. *J Neurovirol* 1999;**5**:32−41.
38. Luster AD. Chemokines-chemotactic cytokines that mediate inflammation. *N Engl J Med* 1998;**338**:436−45.
39. Shoham S, Ebstein RP. The distribution of β-amyloid precursor protein in rat cortex after systemic kainite-induced seizures. *Exp Neurol* 1997;**147**:361−76.
40. Combs CK, Karlo JC, Kao SC, Landreth GE. Beta-amyloid stimulation of microglia and monocytes results in TNF-α-dependent expression of inducible nitric oxide synthase and neuronal apoptosis. *J Neurosci* 2001;**21**:1179−88.
41. Schmidt R, Schmidt H, Curb JD, Masaki K, White LR, Launer LJ. Early inflammation and dementia: a 25-year follow-up of the Honolulu-Asia Aging Study. *Ann Neurol* 2002;**52**:168−74.
42. Li M, Pisalyaput K, Galvan M, Tenner AJ. Macrophage colony stimulatory factor and interferon-γ trigger distinct mechanisms for augmentation of β-amyloid-induced microgliamediated neurotoxicity. *J Neurochem* 2004;**91**:623−33.
43. Tan J, Town T, Paris D, Mori T, Suo Z, Crawford F, et al. Microglial activation resulting from CD40-CD40L interaction after β-amyloid stimulation. *Science* 1999;**286**:2352−5.
44. Webster SD, Park M, Fonseca MI, Tenner AJ. Structural and functional evidence for microglial expression of C1qRP, the C1q receptor that enhances phagocytosis. *J Leukoc Biol* 2000;**67**:109−16.
45. Webster SD, Galvan MD, Ferran E, Garzon-Rodriguez W, Glabe CG, Tenner AJ. Antibody-mediated phagocytosis of the amyloid β-peptide in microglia is differentially modulated by C1q. *J Immunol* 2001;**166**:7496−503.
46. D'Andrea MR. Evidence the immunoglobulin-positive neurons in Alzheimer's disease are dying via the classical antibody-dependent complement pathway. *Am J Alz Dis Other Dementias* 2005;**20**:144.
47. D'Andrea MR. Evidence linking neuronal cell death to autoimmunity in Alzheimer's disease. *Brain Res* 2003;**982**:19−30.
48. D'Andrea MR. Add Alzheimer's disease to the list of autoimmune diseases. *Med Hypothesis* 2005;**64**:458.
49. Fan R, Tenner AJ. Differential regulation of Aβ42-induced neuronal C1q synthesis and microglial activation. *J Neuroinflam* 2005;**2**:1.
50. Lu B, Rutledge BJ, Gu L, Fiorillo J, Lukacs NW. Abnormalities inmonocyte recruitment and cytokine expression in monocyte chemoattractant protein 1-deficient mice. *J Exp Med* 1998;**187**:601−8.

51. Ishizuka K, Kimura T, Igata-yi R, Katsuragi S, Takamatsu J, Miyakawa T. Identification of monocyte chemoattractant protein- 1 in senile plaques and reactive microglia of Alzheimer's disease. *Psychiatry Clin Neurosci* 1997;**51**:135—8.
52. Galimberti D, Schoonenboom N, Scarpini E, Scheltens P. Chemokines in serum and CSF of Alzheimer's disease patients. On behalf of the DIAR Group *Ann Neurol* 2003;**53**:547—8.
53. Sun YX, Minthon L, Wallmark A, Warkentin S, Blennow K, Janciauskiene S. Inflammatory markers in matched plasma and cerebrospinal fluid from patients with Alzheimer's disease. *Dement Geriatr Cogn Disord* 2003;**16**:136—44.
54. Lue L-F, Yan SD, Stern DM, Walker DG. Review. Preventing activation of receptor for advanced glycation endproducts in Alzheimer's disease. *Curr Drug Targets-CNS Neurol Disord* 2005;**4**:249—66.
55. Lue L-F, Rydel R, Brigham EF, Yang LB, Hampel H, Murphy Jr. GM, et al. Inflammatory repertoire of Alzheimer's disease and nondemented elderly microglia in vitro. *Glia* 2001;**35**:72—9.
56. Imai Y, Kohsaka S. Intracellular signaling in M-CSF-induced microglia activation: role of Iba1. *Glia* 2002;**40**:164—74.
57. Takeuchi A, Miyaishi O, Kiuchi K, Isobe KJ. Macrophage colony-stimulating factor is expressed in neuron and microglia after focal brain injury. *Neurosci Res* 2001; **65**:38—44.
58. De Jonge WJ, Ulloa L. Review. The α7 nicotinic acetylcholine receptor as a pharmacological target for inflammation. *Br J Pharmacol* 2007;**151**:915—29.
59. Salhia B, Angelov L, Roncari L, Wu X, Shannon P, Guhaa A. Expression of vascular endothelial growth factor by reactive astrocytes and associated neoangiogenesis. *Brain Res* 2000;**883**:87—97.
60. Norenberg MD. Astrocyte response to CNS injury. *J Neuropathol Exp Neurol* 1994;**53**:213—20.
61. Pike CJ, Cummings BJ, Monzavi R, Cotman CW. Beta-amyloid-induced changes in cultured astrocytes parallel reactive astrocytosis associated with senile plaques in Alzheimer's disease. *Neuroscience* 1994;**63**:517—31.
62. Kato M, Saito H, Abe K. Nanomolar amyloid β protein-induced inhibition of cellular redox activity in cultured astrocytes. *J Neurochem* 1997;**68**:1889—95.
63. Canning DR, McKeon RJ, DeWitt DA, Perry G, Wujek JR, Frederickson RCA, et al. Beta-amyloid of Alzheimer's disease induces reactive gliosis that inhibits axonal growth. *Exp Neurol* 1993;**124**:289—98.
64. Abe K, Kato M, Saito H. Human amylin mimics amyloid β protein-induced reactive gliosis and inhibition of cellular redox activity in cultured astrocytes. *Brain Res* 1997;**762**:285—8.
65. Abe K, Hisatomi R, Misawa M. Amyloid peptide specifically promotes phosphorylation and nuclear translocation of the extracellular signal-regulated kinase in cultured rat cortical astrocytes. *J Phaemacol* 2003;**93**:272—8.
66. Merlini M, Meyer EP, Ulmann-Schuler A, Nitsch RM. Vascular β-amyloid and early astrocyte alterations impair cerebrovascular function and cerebral metabolism in transgenic arcAb mice. *Acta Neuropathol* 2011;**122**:293—311.
67. Nagele RG, D'Andrea MR, Lee H, Venkataraman V, Wang H-Y. Astrocytes accumulate Aβ42 and give rise to astrocytic amyloid plaques in Alzheimer disease brains. *Brain Res* 2003;**971**:197—209.
68. Yua W-F, Guana Z-Z, Bogdanovic N, Nordberg A. High selective expression of α7 nicotinic receptors on astrocytes in the brains of patients with sporadic Alzheimer's disease and patients carrying Swedish APP 670/671 mutation: a possible association with neuritic plaques. *Exp Neurol* 2005;**192**:215—25.

69. Li C, Zhao R, Gao K, Wei Z, Yin MY, Lau LT, et al. Review. Astrocytes: implications for neuroinflammatory pathogenesis of Alzheimer's disease. *Curr Alzheimer Res* 2011;**8** (1):67−80.
70. Young MJ, Lee RK, Jhaveri S, Wurtman RJ. Intracellular and cell-surface distribution of amyloid precursor protein in cortical astrocytes. *Brain Res Bull* 1999;**50**(1):2732.
71. Zhao J, O'Connor T, Vassar R. Review. The contribution of activated astrocytes to Aβ production: Implications for Alzheimer's disease pathogenesis. *J Neuroinflam* 2011;**8**:150.
72. Busciglio J, Gabuzda BH, Matsudaira P, Yanker BA. Generation of β-amyloid in the secretory pathway in neuronal and nonneuronal cells. *PNAS* 1993;**90**:2092−6.
73. Siman R, Card J, Nelson R, Davis L. Expression of β-amyloid precursor protein in reactive astrocytes following neuronal damage. *Neuron* 1989;**3**:275−85.
74. Tang BL. Neuronal protein trafficking associated with Alzheimer disease: from APP and BACE1 to glutamate receptors. *Cell Adh Migr* 2009;**3**:118−28.
75. Yu Y, He J, Zhang Y, Luo H, Zhu S, Yang Y, et al. Increased hippocampal neurogenesis in the progressive stage of Alzheimer's disease phenotype in an APP/PS1 double transgenic mouse model. *Hippocampus* 2009;**19**:1247−53.
76. Otth C, Concha, Arendt T, Stieler J, Schliebs R, Gonzalez-Billault C, et al. AβPP induces cdk5-dependent tau hyperphosphorylation in transgenic mice Tg2576. *J Alzheimers Dis* 2002;**4**:417−30.
77. Morales I, Farías G, Maccioni RB. Neuroimmunomodulation in the pathogenesis of Alzheimer's disease. *Neuroimmunomodulation* 2010;**17**(3):202−4.
78. Kitazawa M, Cheng D, Tsukamoto MR, Koike MA, Wes PD, Vasilevko V, et al. Blocking IL-1 signaling rescues cognition, attenuates tau pathology, and restores neuronal β-catenin pathway function in an Alzheimer's disease model. *J Immunol* 2011;**187** (12):6539−49.
79. Ghosh S, Wu MD, Shaftel SS, Kyrkanides S, LaFerla FM, Olschowka JA, et al. Sustained interleukin-1β overexpression exacerbates tau pathology despite reduced amyloid burden in an Alzheimer's mouse model. *J Neurosci* 2013;**33**(11):5053−64.

Consequences of Intracellular Amyloid in Vascular System

Aβ can impact other cells besides neurons. After a very brief overview of the pathological vascular role of Aβ in AD, this chapter will present the normal and pathological consequences of intracellular Aβ in the cell types of the vascular system. This chapter will present additional information about the pathological aspects of the blood–brain barrier (BBB) in the context of AD.

Intracellular Consequences of Amyloid in Alzheimer's Disease.
DOI: http://dx.doi.org/10.1016/B978-0-12-804256-4.00008-5

155

AD AS A VASCULAR DISEASE

Overwhelming evidence shows that sporadic AD is a vascular disorder.[1-3] Most cases of AD are accompanied by cerebrovascular pathologies, including cerebral amyloid angiopathy (CAA), degeneration of the vascular cells (eg, smooth muscle cells, endothelial cells, and pericytes), and hypoperfusion.[4-9]

Physiological functions of the cerebral vasculature are compromised early, possibly even before the onset of AD pathology (including the onset of CAA) suggesting that the initial pathological event in AD may originate in the vascular system.[10,11] The presence of cerebral hypoperfusion precedes hypometabolism, cognitive decline, and neurodegeneration in patients with MCI and in transgenic mouse models of AD, making regional brain microvascular abnormalities as one of the most upstream pathological factors in AD.[1,12,13]

Due to the cerebrovascular pathologies, brain cells in AD severely suffer from hypoperfusion of glucose and oxygen in the early phases of both the sporadic and familial forms of AD. The history and coexistence of ischemic stroke increase the risk of AD and the use of medicines designed to improve cerebral perfusion, and reduce the symptoms or progression of AD.[1-3,14-16] These data provide additional evidence that the pathological changes in the vascular system could initiate AD.

The frequent appearance of white matter lesions in AD patients are thought to be predominantly due to vascular lesions that may either be due to amyloid perivascular deposits or coexistent atherosclerosis.[17,18]

Some suggest that the attention of the classical pathological markers of extracellular plaques and NFTs in AD may mask the primary event of the vascular pathology.[19,20] While vascular and metabolic disorders including atherosclerosis and diabetes can strongly impair cerebral circulation, neurotoxic sAβ species can also impact the cerebral vasculature perhaps before the symptoms of cognitive impairment are presented.[21,22] This presumably happens at early and pre-pathological stages of Aβ pathology itself. Through the induction of microvascular cell dysfunction and exacerbation of inflammation-related processes, pre-fibrillar Aβ can alter cerebral blood flow and influence the permeability properties of the BBB, both of which are critical for the maintenance of brain homeostasis and ultimately normal neuronal function.[23,24]

Risk Factors

It is quite compelling that based on epidemiological studies, practically all of the risk factors for AD have a vascular component. Other vascular-related risk factors that significantly increase the risk of

AD include hypertension, atherosclerosis, head trauma, obesity, type II diabetes mellitus, hypercholesterolemia (total cholesterol), smoking, oxidative stress, and cardiac disease.[1–4,25–28] The degree to which the vascular risk factors contribute to AD are also influenced by genetic factors such as *APOE*, which has a role in both AD and vascular diseases.[24,29] Among all of these factors, age is still considered the greatest risk factor for late-onset AD. Most of these vascular pathologies cause cerebral ischemia that is commonly present in AD.[1–4,30] Autopsy studies also show the presence of vascular abnormalities including endothelial damage associated with microangiopathy in brain tissue from almost all AD patients in the advanced stages.[31]

Cerebral Amyloid Angiopathy

CAA is associated with a plethora of abnormalities and malfunctioning of the cerebrovasculature including ruptures of the vessel walls causing microhemorrhages and lesions of the BBB. These result in inflammation, vascular edema, and uncontrolled influx of peripheral blood components into the brain parenchyma.[23,32–35] As a result of CAA, vascular basement membranes thicken and smooth muscle cells degenerate, leading to impaired cerebral blood flow that is associated with cognitive decline.[25,32,33,36] These vascular pathologies are associated with the deposition of Aβ within the walls of the leptomeninges and parenchymal arteries, arterioles walls, and capillaries. These depositions lead to concomitant thickening of arteriole walls and formation of microaneurysms with degeneration of vessel wall components. The majority of the amyloid deposits in the walls of the larger vessels in AD patients was not associated with a chronic inflammatory response, in contrast to microcapillary amyloid angiopathy, which does incite an inflammatory response.[37]

CAA is a critical factor in the pathogenesis of AD.[36] As many as 80% of AD patients and 10–40% of elderly without AD have CAA.[38,39] Another report indicated nearly 98% AD patients have CAA, and 75% of these patients are rated as severe CAA.[36] Down syndrome patients also have prominent CAA due to their excessive production of Aβ.[40,41] The deposition of Aβ on the cerebrovascular walls in patients with CAA is also an important pathological feature of familial AD and is present in more than 90% of sporadic AD patients.[42] Cerebral vessel fractions derived from late-onset AD patients with CAA contained fivefold more Aβ than AD patients without CAA.

The risk of CAA increases in carriers of the *APOE* ε4 genotype.[43] One of the proposed mechanisms leading to CAA begins with internalization and accumulation of Aβ–ApoE complexes in cerebral smooth muscle cells. Such a model could explain the preferential localization of CAA to the outer and middle layers of cortical and leptomeningeal arterioles, while indicating a mechanism by which the *APOE* genotype might determine the risk of CAA.[43]

Vascular Dementia

Vascular dementia (VaD) is the second most common cause of dementia in the elderly after AD. VaD is defined as loss of cognitive function resulting from ischemic, hypoperfusive, or hemorrhagic brain lesions due to cerebrovascular disease or cardiovascular pathology. One distinguishing contrast between VaD and AD is the onset of the disease: VaD can occur suddenly, while AD is a gradual but steady decline. Patients with VaD also exhibit executive function impairment rather than the memory loss that characterizes AD.[44]

However, VaD and AD share a number of risk factors, clinical cognitive impairment symptoms, and cerebrovascular lesions.[1,44] In a study, 2800 patients with hypertension were followed for about 4 years.[45] Compared to the controls, those patients who received long-term antihypertensive therapy reduced their risk of dementia by 55%, from 7.4 to 3.3 cases per 1000 patient-years. Of interest is the observation that treatment with nitrendipine, a calcium channel blocker, prevented AD and poststroke VaD suggesting that calcium channel blockers, by improving calcium dysregulation, may have neuroprotective effects in VaD and AD beyond blood pressure control.[46,47]

Type 2 Diabetes

Type 2 diabetic patients have twice the risk of developing AD as nondiabetic patients.[48] Defective insulin signaling in diabetic patients can lead to synaptic pathology.[49] Diabetes is characterized by hyperglycemia with several macrovascular (coronary artery disease, peripheral arterial disease, and stroke) and microvascular (diabetic nephropathy, neuropathy, and retinopathy) complications.[50] As in the AD neurons and other cells, Aβ can accumulate in the pancreatic insulin secreting β-cells leading to their degeneration as well.[49] Preclinical models of diabetes show the development of AD and VaD.[51,52] In one study, hypoinsulinemic diabetes was induced with streptozotocin (STZ) in transgenic (APP/PS1) mice at 18 weeks of age, well before the typical presence of AD pathology in these mice. The STZ-induced episodic and working memory impairment was significantly worsened in transgenic mice as compared with the nontreated transgenic mice. The presence of hemorrhages was significantly higher in treated transgenic mice, and although pericytes and endothelium were only partially affected, BBB alterations may underlie observed pathological features.[51,52]

SOURCE OF THE VASCULAR-LADENED Aβ

Aβ in the vessels can originate from the brain, from systemic circulation (ie, peripheral cells), and/or from the vascular cells themselves.[53] Aβ is present in all of these areas: the cerebral spinal and brain interstitial fluids, the peripheral cells and plasma, and in all of the vascular cells including the endothelial cells, smooth muscle cells, adventitial cells, brain pericytes, and perivascular cells.[9,54] All are implicated as the origin of Aβ in vascular amyloid.

Brain-Derived Aβ

Although the developments of CAA and parenchymal Aβ deposition might occur independently, neuronal Aβ is associated with the basement membrane around capillaries, and is transported out of the brain through perivascular drainage pathways.[55,56] The Aβ produced in the brain could flow with interstitial fluid out of the brain and into the vessels where the Aβ becomes lodged in the blood vessels walls becoming an anchor for additional Aβ deposition.[57] The presence of CAA in transgenic mouse models of AD also supports this possible mechanism.[58,59]

Local production of increased levels of Aβ does not seem likely even though Aβ detection is detected in the many cells of the vascular system.[53] However, the possibility of the brain contributing large amounts of Aβ to local areas of CAA also seemed unlikely, leaving the most logical source of the Aβ deposition in the cerebrovasculature as the vascular system.[53,60]

Peripherally Derived Aβ

Various peripheral cells contribute to plasma Aβ concentrations (see Chapter 2). However, only the role of the platelets is discussed here in relation to intracellular Aβ in vascular cells. Vascular injury could be one of the first vascular pathological events in CAA and AD that leads to platelet activation and adhesion to the vascular wall (Fig. 8.1). Activated platelets contribute to more than 95% of circulating Aβ, which in turn activates platelets and results in the vicious cycle of Aβ overproduction in the damaged vessel (see Chapter 1).[36,61] The uncontrolled activation of platelets leads to a chronic inflammatory reaction by secretion of chemokines.[36] The interaction of these biological response modulators with platelets, endothelial cells, and leukocytes establishes a localized inflammatory response that can also contribute to CAA. Activated platelets are structurally abnormal and resistant to degradation in the

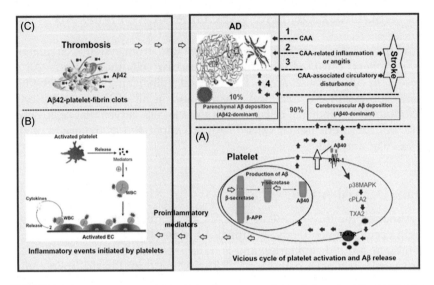

FIGURE 8.1   Contribution of blood platelets to vascular pathology in AD. (A) Platelet-originated Aβ40 vicious cycle; (B) inflammatory events initiated by activated platelets; (C) Aβ42-platelet-fibrin clot-induced thrombus formation. *AD*, Alzheimer's disease; *Aβ40*, β-amyloid40; *APP*, amyloid precursor protein; *CAA*, cerebral amyloid angiopathy; *cPLA*, cytosolic phospholipase A; *EC*, endothelial cell; *PAR*, protease-activated receptor; *TXA*, thromboxane A; *TXA2R*, thromboxane A2 receptor; *WBC*, white blood cell. *Source: Used with permission from J Blood Med 2013;4:141−7.*

presence of Aβ42. Thus, a recent proposal suggests that targeting blood platelets may provide a new avenue for anti-AD therapy.

Metabolic dysfunction can lead to abnormal Aβ production in AD platelets, which can lead to alterations in APP metabolism.[8,62,63] Other studies, performed on platelets obtained from moderate to severe AD patients, reported a plethora of abnormalities: cytoskeletal abnormalities, changes in the degree of activation, cytochrome oxidase deficiency, abnormal cytoplasmic calcium fluxes, increased platelet level of COX-2, abnormal glutamate transporter activity, decreased phospholipase C activity, and increased cytosolic protein kinase C levels.[63−67] In addition, AD patients also show a differential level of platelet APP forms.[9]

Alterations in the levels of α-secretase ADAM10 and in the enzymatic activities of α- and β-secretase are present in platelets of patients with AD, which is consistent with increased processing through the amyloidogenic pathway.[68] APP-cleaving enzyme activity is increased by 24% in platelet membranes of patients with MCI and is increased by 17% in those with AD. The increased production of Aβ could certainly account for the Aβ detected in the vessels of patients with CAA, VaD, and AD.

Locally Derived Aβ

The presence of Aβ in capillaries and in the adventitial cells suggests that its source cannot exclusively originate from smooth muscle cells.[42,54] Although Aβ mRNA expression is detected in smooth muscle cells and endothelial cells, many previous in situ hybridization studies reported absence of Aβ mRNA expression in cerebral blood vessel walls.[69–71] A subsequent study confirmed the presence of APP mRNA at areas of amyloid formation in the vessels suggesting that other vascular cells such as endothelial, adventitial cells, and pericytes can also contribute to cerebrovascular amyloidosis.[54] The production of Aβ in all vascular sites at which amyloid deposition can occur suggests an important contribution of locally derived Aβ to cerebrovascular amyloid deposits.

PATHOLOGICAL ASPECTS OF THE BBB

The BBB regulates the exchanges of Aβ peptide between the blood and the brain, and a deregulation of these exchanges can account for the accumulation of Aβ observed in the AD brain.[72] The disruption of the BBB is associated with cognitive impairment, which may represent one of the earliest pathological events in AD prior to the onset of clinical dementia.[5]

In vivo studies that monitor the ratio of CSF albumin to serum albumin in AD patients show the frequent occurrences of BBB dysfunction that correlates to the severity of the dementia.[73,74] In addition, BBB permeability precedes amyloid plaque formation in Tg2576 mice.[73,75]

One of the anatomic areas particularly vulnerable to BBB compromise in AD is the arterioles, which have a higher pressure than in capillaries and venules.[76] Immunohistochemical analyses of postmortem human brains showed vascular leakage based on the presence of plasma components that were common in AD brains and were primarily associated with arterioles (Figs. 2.2 and 2.3) (see Chapter 2).

The Endothelial Cells of the BBB

Endothelial cells line the inside of the vessels, and the gaps between them vary depending on the area in the body. The brain microvascular endothelial cells along with the pericytes form a metabolic and physical barrier that separates the rest of the body from the brain. These cells also help to maintain cerebral homeostasis and mediate delivery of glucose and other nutrients from the blood into the brain. Endothelial cells also help to mediate the clearance of toxic metabolic by-products from the CNS back to the circulation.[22,77–79] These cerebrovascular endothelial cells are structurally different from all other endothelial cells in the body due to the presence of tight junctional complexes limiting nonspecific entry of blood components into the brain.[80,81] During pathological conditions of the brain, a faulty endothelium can lose its barrier function, which can lead to disturbances in the brain.[82]

In AD, intracellular Aβ is present in vascular endothelial cells, and like other cell types (eg, neurons, smooth muscle cells, and pericytes), excessive intracellular accumulation leads to degeneration.[5,83,84] Aβ can cause endothelial cells to lose function leading to increased vasoconstriction and decreased endothelium-dependent vasodilatation.[85] Aβ can affect cell-to-cell and cell-to-matrix interactions that can increase the adhesiveness leukocytes to the endothelium.[85] Aβ42 can disturb tight-junction protein complexes leading to a dysfunctional BBB.[86,87] Dysfunction of the endothelial cells also correlates with AD severity.[88]

These endothelial abnormalities are also present in predemented patients. Levels of thrombomodulin and other vascular adhesion molecules are increased in patients with mild AD and in subjects with MCI.[88]

In vitro studies using endothelial cells from aortic or pulmonary origins and from cell lines show that Aβ is toxic.[89,90] Furthermore, these Aβ can cause cerebral endothelial cells to become apoptotic.[91]

Vascular endothelial cells also express α7 receptor protein and mRNA. As previously described, Aβ has an extremely high affinity to the α7 receptors and suggests a role in typical endothelial functions in the processes of angiogenesis, inflammation, and atherogenesis (see Chapter 2).[92,93] The presence of the α7 receptors can account for the presence of intracellular Aβ in the endothelial cells through the processes of receptor-mediated endocytosis (see Chapter 2). Although the natural expression of the α7 receptors is extremely low (∼50 fmol/mg protein), in vitro experiments show that agonists such as nicotine can significantly increase α7 receptor levels (∼300-fold).[93] The upregulation of the α7 receptors can easily explain the overaccumulation of intracellular Aβ in virtually any cell type with these receptors, including the vascular cells and neurons.

The Pericytes

Amyloid-laden capillaries in the AD brain are also marked by pericytes loss and degeneration that are observed in CAA (Fig. 8.2).[5,81] Pericytes are contractile cells that wrap around the endothelial cells of capillaries and venules throughout the body (Fig. 2.1) (see Chapter 2). The role of pericytes in the BBB is to help maintain the barrier function along with

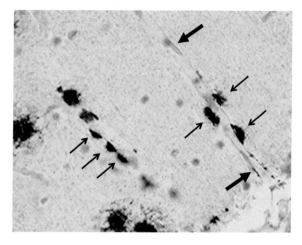

FIGURE 8.2 Presence of Aβ42 in the Alzheimer's brain shows intracellular Aβ in apparent vascular pericytes (*small arrows*), as well as in several nearby plaques (left and bottom). Aβ42 immunoreactivity was also detected in one (*big arrow*, bottom) of the capillary endothelial cell, but not in the other (*big arrow*, top). Pericytes were reported to be more associated with amyloid deposition than with endothelial cells. *Aβ42*, amyloid-β42. *Source: Used with permission from Slidomics, LLC.*

the endothelial and astrocytic cells. Like smooth muscle cells, pericytes can also change the rate of microcirculation and cerebral blood flow at the capillary level.[5] Pericytes isolated from brain microvessels can also degrade Aβ.[53]

In chronic BBB disruption models, pericytes can also remove toxic circulating plasma proteins that are normally excluded from the brain; these include immunoglobulins, fibrin, and albumin.[5] Pericytes can phagocytose cellular debris in models of acute brain injury.

In vitro studies of bovine retinal capillary pericytes show that low concentrations (2–20 μM) of Aβ cause marked ultrastructural changes such as cell body shrinkage, retraction of processes, and disruption of the intracellular actin structure.[94] These data suggest that together with other vascular cells, pericytes could be the target of vascular damage during processes involving amyloid accumulation.

Aβ can internalize in pericytes through the low-density lipoprotein receptor-related protein 1 (LRP1), which is strongly expressed in pericytes.[72] LRP1, also known as the clearance receptor, internalizes Aβ for lysosomal degradation.[95,96] High concentrations and prolonged exposure of Aβ can kill pericytes in vitro. Accumulation of intracellular Aβ in the pericytes as well as reductions in pericyte cell numbers were reported in transgenic APP mice, which is sufficient to disrupt BBB function and lead to hypoperfusion (Fig. 8.3).[95] In spite of these reports

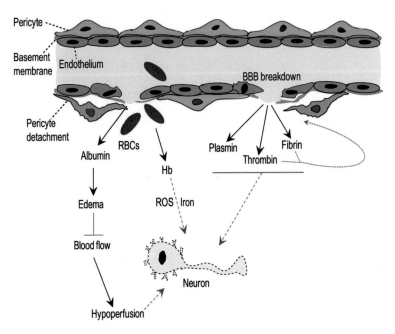

FIGURE 8.3   Neurovascular functions of central nervous system pericytes. Pericyte leads to a chronic BBB disruption and vascular-mediated secondary neuronal injury and degeneration. Pericyte loss and/or degeneration represent(s) an important cellular source of hit 1 vascular injury in Alzheimer's disease. Pericyte degeneration leads to BBB disruption and unrestricted entry and accumulation of blood-derived products in brain including erythrocyte-derived hemoglobin and plasma-derived proteins such as albumin, plasmin, thrombin, fibrin, and immunoglobulins. Plasmin and thrombin have direct neurotoxic properties, whereas fibrin accelerates neurovascular injury. Brain degradation of hemoglobin liberates free iron which catalyzes the formation of ROS leading to further injury. Albumin increases oncotic pressure resulting in edema, microvascular compression, and reduced blood flow. Pericyte loss also leads to endothelial cell death and microvascular regression leading to additional simultaneous reductions in blood flow. In mouse models, vascular injury in absence of Aβ as a result of pericyte loss is sufficient for neurodegeneration. *Aβ*, amyloid-β; *BBB*, blood−brain barrier; *ROS*, reactive oxygen species. *Source: Used with permission from* Brain Pathol 2014;**24**(4):371−86.

on how Aβ is internalized in pericytes, the source of the Aβ may be from vascular smooth muscle cells (VSMCs) and/or pericytes themseleves.[54,97] Regardless of the source of the intracellular Aβ in the pericytes, when pericytes die, the BBB will become dysfunctional resulting in the unregulated leakage of vascular components into the brain.[5]

The Smooth Muscle Cells

Amyloid angiopathy in the AD brain is often accompanied by smooth muscle cell death.[7,98] The excessive accumulations of intracellular Aβ42 have been reported in VSMCs that may lead to their demise thereby contributing to the breakdown of the BBB.[99,100] Intracellular Aβ disrupted smooth muscle actin leading to cell death.[29]

The function of the VSMCs is to help control perfusion rate by regulating the diameter of the vessels. They exist in a wide range of anatomic locations in the body and are known to have different functions since they have varying patterns of gene expression.[101] As an example, under hypoxic conditions, Aβ uptake in VSMCs was not similar in VSMCs from various embryonic origins.[102]

Abundant Aβ40 and Aβ42 immunoreactivity was detected in VSMCs of the AD brains,[29,100] especially in leptomeningeal arteries and their initial cortical branches, which extended into the microvasculature in the more severe cases of AD.[100] Diffuse, perivascular Aβ plaques were associated primarily with arterioles, like those observed in Fig. 5.4, and demonstrate the presence of local areas of BBB dysfunction.

The Aβ is internalized into the lysosomes in the VSMCs.[43] Excessive intracellular accumulation of Aβ in the VSMCs can lead to their degeneration leaving the Aβ within the extracellular space of the vessel wall thereby contributing to CAA, like that hypothesized about the demise of the neurons.[29,71,103] The intracellular Aβ can be internalized through class A and class B scavenger receptors because these receptors are detected in Aβ-containing endosomes and because Aβ uptake is inhibited by various scavenger receptor antagonists (Fig. 5.4).[83,99,104,105] Supportive experiments show rapid internalization of Aβ in endosomes and lysosomal vesicles in VSMCs in the presence of human serum or CSF.[43] Furthermore, receptor-mediated endocytosis of Aβ was enhanced by chloroquine and blocked by cycloheximide, brefeldin A, and pretreatment with trypsin.[43]

VSMCs can also produce Aβ as evidenced by the expression of Aβ mRNA, and Aβ is able to induce its own production in VSMCs in vitro.[54,106] However, the presence of amyloid in capillaries (areas devoid of smooth muscle cells) rules out the possibility that smooth muscle cells are the only source of Aβ in CAA.[107]

Receptor-Mediated Intracellular Aβ

LRP1 is abundantly expressed in several different cell types in the brain including neurons, glia, and VSMCs.[29,43,96,108−110] LRP1 can bind and internalize Aβ via a lipoprotein pathway involving ApoE.[104,111] As noted, once internalized into the VSMCs, Aβ was subsequently trafficked to the lysosomes that contain a variety of acid hydrolases to degrade the Aβ.[102,103,112] When the lysosomal function is hampered, Aβ can accumulate and aggregate (see Chapters 4, 5).[113]

The suppression of the LRP1 in VSMCs of the brain significantly diminished their ability to uptake and degrade exogenous Aβ.[102]

Endocytosis of Aβ into pericytes and smooth muscle cells is also mediated by RAGE.[53]

The cerebrovascular smooth muscle cells express the α7 receptor, which has high binding affinity for Aβ peptides, especially Aβ42 (see Chapter 2). The distribution of the α7 receptor is not uniform across all VSMCs,[92] which correlates to the studies showing differential uptake of Aβ in VSMCs from a variety of anatomical locations.[100,101] The inconsistent presence of Aβ in the renal circulation distal to the main renal artery may be explained by the absence of the α7 receptor in these anatomic areas.[92]

The detection of intracellular Aβ in VSMCs in the AD tissues suggests that the source of the Aβ may originate from the vascular system perhaps via the platelets and peripheral epithelial tissues (see Chapter 2). Also the dysfunction in the BBB could provide a major source of the Aβ that gradually accumulates in brain VSMCs (see Chapter 2). The presence and abundance of the α7 receptor on VSMCs may facilitate the selective accumulation of Aβ peptides in these cells.[100]

SUMMARY

Overwhelming evidence provided in this chapter, and from many publications well beyond those listed in this book, convincingly shows that vascular pathologies are not only present in AD but may actually be one of the earliest pathological events leading to the disease. It is not clear which group of patients with vascular diseases eventually develop AD, CAA, or VaD, but it is clear that vascular pathology is a prerequisite for AD. All of the cells of the vascular system contribute to the pathology, which appears to begin from intracellular Aβ. Although the source(s) of the Aβ in these cells is/are not clear, the presence of several Aβ receptors provides a mechanism of endocytosis of the intracellular Aβ.

The BBB may be the portal where Aβ enters the AD brain. The collective orchestration of the vascular cells helps to maintain the barrier function, and it is clear that all of these cells are negatively impacted by intracellular levels of Aβ. Loss of BBB function, which is present in most of the AD brains, leads to focal areas of vascular leakage. As previously reviewed, the presence of Aβ-high affinity receptors on neurons suggests that CNS levels of Aβ are highly regulated. However, if vascular-derived Aβ pours into the brain, the neurons continue to over-accumulate Aβ eventually leading to their demise. This mechanism can explain why vascular pathology is an early event, and why cognitive impairment occurs subsequently.

References

1. de la Torre JC. Review. Alzheimer disease as a vascular disorder: nosological evidence. *Stroke* 2002;**33**(4):1152−62.
2. de la Torre JC. Review. Is Alzheimer's disease a neurodegenerative or a vascular disorder? Data, dogma, and dialectics. *Lancet Neurol* 2004;**3**(3):184−90.
3. de la Torre JC. Review. Vascular basis of Alzheimer's pathogenesis. *Ann NY Acad Sci* 2002;**977**:196−215.
4. Yang S-P, Baea D-G, Kang HJ, Gwag BJ, Ghoc YS, Chaea C-B. Co-accumulation of vascular endothelial growth factor with β-amyloid in the brain of patients with Alzheimer's disease. *Neurobiol Aging* 2002;**25**:283−90.
5. Winkler EA, Sagare AP, Zlokovic BV. Review. The pericyte: a forgotten cell type with important implications for Alzheimer's disease? *Brain Pathol* 2014;**24**(4):371−86.
6. Wilhelmus MMM, Otte-Holler I, Davis J, Van Nostrand WE, de Waal RMW, Verbeek MM. Apolipoprotein E genotype regulates amyloid-β cytotoxicity. *J Neurosci* 2005;**25**(14):3621−7.
7. Wisniewski HM, Frackowiak J, Mazur-Kolecka B. In vitro production of β-amyloid in smooth muscle cells isolated from amyloid angiopathy-affected vessels. *Neurosci Lett* 1995;**183**(12):1203.
8. Borroni B, Akkawi N, Martini G, Colciaghi F, Prometti P, Rozzini L, et al. Microvascular damage and platelet abnormalities in early Alzheimer's disease. *J Neurological Sci* 2002;**203**:189−93.
9. Cattabeni F, Colciaghi F, Di Luca M. Platelets provide human tissue to unravel pathogenic mechanisms of Alzheimer disease. *Prog Neuro Psychopharmacol Biol Psychiatr* 2004;**28**:763−70.
10. de la Torre JC. Review. The vascular hypothesis of Alzheimer's disease: bench to bedside and beyond. *Karg Neurodeg Dis* 2010;**7**:116−21.
11. Vagnucci Jr AH, Li DW. Alzheimer's disease and angiogenesis. *Lancet* 2003;**361**:605−8.
12. Bourasset F, Ouellet M, Tremblay C, et al. Reduction of the cerebrovascular volume in a transgenic mouse model of Alzheimer's disease. *Neuropharmacology* 2009;**56**:808−13.
13. Dai W, Lopez OL, Carmichael OT, Becket JT, Kuller LH, Gach HM. Mild cognitive impairment and Alzheimer disease: patterns of altered cerebral blood flow at MR imaging. *Radiology* 2009;**250**:856−66.
14. Kalaria RN. Review. The role of cerebral ischemia in Alzheimer's disease. *Neurobiol Aging* 2000;**21**(2):321−30.
15. Kokmen E, Whisnant JP, O'Fallon WM, Chu C-P, Beard CM. Dementia after ischemic stroke: a population-based study in Rochester, Minnesota (1960−1984). *Neurology* 1996;**46**(1):154−9.
16. Snowdon DA, Greiner LH, Mortimer JA, Riley KP, Greiner PA, Markesbery WR. Brain infarction and the clinical expression of Alzheimer disease. The Nun Study. *JAMA* 1997;**277**(10):813−17.
17. Brun A, Englund E. Brain changes in dementia of Alzheimer's type relevant to neuroimaging diagnostic methods. *Prog Neuropsychopharmacol Biol Psychiatr* 1986;**10**:297−308.
18. Rezek DL, Morris JC, Fulling KH, Gado MH. Periventricular white matter lucencies in senile dementia of the Alzheimer type and in normal aging. *Neurology* 1987;**37**:1365−8.
19. Kovari E, Gold G, Herrmann FR, et al. Lewy body densities in the entorhinal and anterior cingulate cortex predict cognitive deficits in Parkinson's disease. *Acta Neuropathol* 2003;**106**:83−8.

20. Gold G, Kovari E, Corte G, et al. Clinical validity of Aβ-protein deposition staging in brain aging and Alzheimer disease. *J Neuropathol Exp Neurol* 2001;**60**:946–52.
21. Kalaria RN. Review. Neurodegenerative disease: diabetes, microvascular pathology and Alzheimer disease. *Nat Rev Neurol* 2009;**5**:305–6.
22. Frackowiak J, Zoltowska A, Wisniewski HM. Non-fibrillar β-amyloid protein is associated with smooth muscle cells of vessel walls in Alzheimer disease. *J Neuropathol Exp Neurol* 1994;**53**:637–45.
23. Zlokovic BV. Review. The blood–brain barrier in health and chronic neurodegenerative disorders. *Neuron* 2008;**57**:178–201.
24. Ghiso J, Fossati S, Rostagno A. Amyloidosis associated with cerebral amyloid angiopathy: cell signaling pathways elicited in cerebral endothelial cells. *J Alz Dis* 2014;**42**(3): S167–76.
25. Dickstein DL, Walsh J, Brautigam H, Stockton Jr SD, Gandy S, Hof PR. Role of vascular risk factors and vascular dysfunction in Alzheimer's disease. *Mt Sinai J Med* 2010;**77**:82–102.
26. Guskiewicz KM, Marshall SW, Bailes J, et al. Association between recurrent concussion and late-life cognitive impairment in retired professional football players. *Neurosurgery* 2005;**57**:719–26.
27. Hölscher C. Diabetes as a risk factor for Alzheimer's disease: insulin signalling impairment in the brain as an alternative model of Alzheimer's disease. *Biochem Soc Trans* 2011;**39**:891–7.
28. Chen J-H, Lin K-P, Chen Y-C. Risk factors for dementia. *J Formos Med Assoc* 2009;**108**:754–64.
29. Ruzali WA, Kehoe PG, Love S. Influence of LRP-1 and apolipoprotein E on amyloid-β uptake and toxicity to cerebrovascular smooth muscle cells. *J Alz Dis* 2013;**33** (1):95–110.
30. Sondell M, Lundborg G, Kanje M. Vascular endothelial growth factor has neurotropic activity and stimulates axonal growth, enhancing cell survival and Schwann cell proliferation in the peripheral nervous system. *J Neurosci* 1999;**19** (14):5731–40.
31. Kalaria RN, Hedera P. Beta-amyloid vasoactivity in Alzheimer's disease. *Lancet* 1996;**347**:1492–3.
32. Farkas E, Luiten PGM. Cerebral microvascular pathology in aging and Alzheimer's disease. *Prog Neurobiol* 2001;**64**:575–611.
33. Pezzini A, Del Zotto E, Volonghi I, Giossi A, Costa P, Padovani A. Cerebral amyloid angiopathy: a common cause of cerebral hemorrhage. *Curr Med Chem* 2009;**16**:2498–513.
34. Smith EE, Greenberg SM. Review. Beta-amyloid, blood vessels, and brain function. *Stroke* 2009;**40**:2601–6.
35. Thal DR, Griffin WST, de Vos RAI, Ghebremedhin E. Cerebral amyloid angiopathy and its relationship to Alzheimer's disease. *Acta Neuropathol* 2008;**115**:599–609.
36. Zhang W, Huang W, Jing F. Contribution of blood platelets to vascular pathology in Alzheimer's disease. *J Blood Med* 2013;**7**:141–7.
37. Rozemiller AJ, van Gool WA, Eikelenboom P. The neuroinflammatory response in plaques and amyloid angiopathy in Alzheimer's disease: therapeutic implications. *Curr Drug Targets CNS Neurol Disord* 2005;**4**(3):223–33.
38. Jellinger KA. Prevalence and impact of cerebrovascular lesions in Alzheimer and Lewy body diseases. *Neurodegener Dis* 2010;**7**:112–15.
39. Viswanathan A, Greenberg SM. Cerebral amyloid angiopathy in the elderly. *Ann Neurol* 2011;**70**:871–80.

40. Rovelet-Lecrux A, Hannequin D, Raux G, Le Meur N, Laquerrière A, Vital A, et al. APP locus duplication causes autosomal dominant early-onset Alzheimer disease with cerebral amyloid angiopathy. *Nat Genet* 2006;**38**:24–6.

41. Sleegers K, Brouwers N, Gijselinck I, Theuns J, Goossens D, Wauters J, et al. APP duplication is sufficient to cause early onset Alzheimer's dementia with cerebral amyloid angiopathy. *Brain* 2006;**129**:2977–83.

42. Maat-Schieman MLC, van Duinen SG, Bornebroek M, Haan J, Roos RAC. Review. Hereditary cerebral hemorrhage with amyloidosis—Dutch type (HCHWA-D): II—A review of histopathological aspects. *Brain Pathol* 1996;**6**:115–20.

43. Urmoneit B, Prikulis I, Wihl G, D'Urso D, Frank R, Heeren J, et al. Cerebrovascular smooth muscle cells internalize Alzheimer amyloid β protein via a lipoprotein pathway: implications for cerebral amyloid angiopathy. *Lab Invest* 1997;**77**(2):157–66.

44. Román GC. Review. Vascular dementia: distinguishing characteristics, treatment, and prevention. *JAGS* 2003;**51**:S296–304.

45. Forette F, Seux M-L, Staessen JA, et al. The prevention of dementia with antihypertensive treatment: new evidence from the Systolic Hypertension in Europe (SYST-EUR) Study. *arch Intern Med* 2002;**162**:2046–52.

46. Morich FJ, Bieber F, Lewis JM, et al. Nimodipine in the treatment of probable Alzheimer's disease: results of two multicentre trials. *Clin Drug Invest* 1996;**11**:185–95.

47. Pantoni L, Rossi R, Inzitari D, et al. Efficacy and safety of nimodipine in subcortical vascular dementia: a subgroup analysis of the Scandinavian Multi-Infarct Dementia Trial. *J Neurol Sci* 2000;**175**:124–34.

48. Biessels GJ, Bravenboer B, Gispen WH. Glucose, insulin and the brain: modulation of cognition and synaptic plasticity in health and disease: a preface. *Eur J Pharmacol* 2004;**490**:1–4.

49. Sandhir R, Gupta S. Review. Molecular and biochemical trajectories from diabetes to Alzheimer's disease: a critical appraisal. *World J Diabetes* 2015;**6**(12):1223–42.

50. Klein R. Hyperglycemia and microvascular and macrovascular disease in diabetes. *Diabetes Care* 1995;**18**:258–68.

51. Ramos-Rodriguez JJ, Infante-Garcia C, Galindo-Gonzalez L, Garcia-Molina Y, Lechuga-Sancho A, Garcia-Alloza M. Increased spontaneous central bleeding and cognition impairment in APP/PS1 mice with poorly controlled diabetes mellitus. *Mol Neurobiol* 2015. [Epub ahead of print]

52. Ramos-Rodriguez JJ, Jimenez-Palomares M, Murillo-Carretero MI, Infante-Garcia C, Berrocoso E, Hernandez-Pacho F, et al. Central vascular disease and exacerbated pathology in a mixed model of type 2 diabetes and Alzheimer's disease. *Psychoneuroendocrinology* 2015;**62**:69–79.

53. Lue L-F, Yan SD, Stern DM, Walker DG. Preventing activation of receptor for advanced glycation endproducts in Alzheimer's disease. *Curr Drug Targets CNS Neurol Disord* 2005;**4**:249–66.

54. Natte R, de Boer WI, Maat-Schieman MLC, Baeld HJ, Vinters HV, Roos RAC, et al. Amyloid β precursor protein-mRNA is expressed throughout cerebral vessel walls. *Brain Res* 1999;**828**:179–83.

55. Preston SD, Steart PV, Wilkinson A, Nicoll JA, Weller RO. Capillary and arterial cerebral amyloid angiopathy in Alzheimer's disease: defining the perivascular route for the elimination of amyloid β from the human brain. *Neuropathol Appl Neurobiol* 2003;**29**:106–17.

56. Weller RO, Massey A, Newman TA, Hutchings M, Kuo YM, Roher AE. Cerebral amyloid angiopathy: amyloid β accumulates in putative interstitial fluid drainage pathways in Alzheimer's disease. *Am J Pathol* 1998;**153**:725–33.

57. Soffer D. Review. Cerebral amyloid angiopathy—a disease or age-related condition. *Isr Med Assoc J* 2006;**8**:803—6.
58. Calhoun ME, Burgermeister P, Phinney AL, Stalder M, Tolnay M, Wiederhold KH, et al. Neuronal overexpression of mutant amyloid precursor protein results in prominent deposition of cerebrovascular amyloid. *PNAS* 1999;**96**:14088—93.
59. Van Dorpe J, Smeijers L, Dewachter I, Nuyens D, Spittaels K, Van Den HC, et al. Prominent cerebral amyloid angiopathy in transgenic mice overexpressing the London mutant of human APP in neurons. *Am J Pathol* 2000;**157**:1283—98.
60. Kuo Y-M, Emmerling MR, Lampert HC, Hempelman SR, Kokjohn TA, Woods AS, et al. High levels of circulating Aβ42 are sequestered by plasma proteins in Alzheimer's disease. *Biochem Biophys Res Commun* 1999;**257**:787—91.
61. Chen M, Inestrosa NC, Ross GS, Fernandez HL. Platelets are the primary source of amyloid β-peptide in human blood. *Biochem Biophys Res Commun* 1995;**213**(1):96—103.
62. Zubenko GS, Wusylko M, Cohen BM, Boller F, Teply I. Family study of platelet membrane fluidity in Alzheimer's disease. *Science* 1987;**238**:539—42.
63. Sevush S, Jy W, Horstman LL, et al. Platelet activation in Alzheimer disease. *Arch Neurol* 1998;**55**(4):530—6.
64. Ciabattoni G, Porreca E, Di Febbo C, et al. Determinants of platelet activation in Alzheimer's disease. *Neurobiol Aging* 2007;**28**(3):336—42.
65. Davies TA, Long HJ, Tibbles HE, Sgro KR, Wells JM, Rathbun WH, et al. Moderate and advanced Alzheimer's patients exhibit platelet activation differences. *Neurobiol Aging* 1997;**18**:155—62.
66. Sage SO, Ronk TJ. Kinetics of changes in intracellular calcium concentration in Fura-2-loaded human platelets. *J Biol Chem* 1987;**262**:16364—9.
67. Matsushima H, Shimohama S, Fujimoto S, Takenawa T, Kimura J. Reduction of platelet phospholipase C activity in patients with Alzheimer's disease. *Alzheimer Dis Assoc Disord* 1995;**9**:213—17.
68. Evin G, Li Q-X. Platelets and Alzheimer's disease: potential of APP as a biomarker. *World J Psychiatr* 2012;**2**(6):102—13.
69. Takahashi RH, Sawa H, Takada A, Kitabatake A, Nagashima K. Expression of amyloid precursor protein mRNA in vascular smooth muscle cells of the human brain. *Neuropathology* 1997;**17**:11—14.
70. Schmechel DE, Goldgaber D, Burkhart DS, Gilbert JR, Gajdusek DC, Roses AD. Cellular localization of messenger RNA encoding amyloid-β-protein in normal tissue and in Alzheimer disease. *Alzheimer Dis Assoc Disord* 1998;**2**:96—111.
71. Tanzi RE, Wenniger JJ, Hyman BT. Cellular specificity and regional distribution of amyloid β protein precursor alternative transcripts are unaltered in Alzheimer hippocampal formation. *Mol Brain Res* 1993;**18**:246—52.
72. Candela P, Saint-Pol J, Kuntz M, Boucau MC, Lamartiniere Y, Gosselet F, et al. In vitro discrimination of the role of LRP1 at the BBB cellular level: focus on brain capillary endothelial cells and brain pericytes. *Brain Res* 2015;**1594**:15—26.
73. Skoog I, Wallin A, Fredman P, Hesse C, Aevarsson O, Karlsson I, et al. A population study on blood—brain barrier function in 85-year-olds: relation to Alzheimer's disease and vascular dementia. *Neurology* 1998;**50**:966—71.
74. Wada H. Blood—brain barrier permeability of the demented elderly as studied by cerebrospinal fluid—serum albumin ratio. *Intern Med* 1998;**37**:509—13.
75. Ujiie M, Dickstein DL, Carlow DA, Jefferies WA. Blood—brain barrier permeability precedes senile plaque formation in an Alzheimer disease model. *Microcirculation* 2003;**10**:463—70.
76. Clifford PM, Zarrabia S, Siua G, Kosciuk MC, Venkataramanb V, D'Andrea MR, et al. Aβ peptides can enter the brain through a defective blood—brain barrier and bind selectively to neurons. *Brain Res* 2007;**1142**:223—36.

77. Zlokovic BV. Review. Neurovascular pathways to neurodegeneration in Alzheimer's disease and other disorders. *Nat Rev Neurosci* 2011;**12**(12):723–38.
78. Andus KL, Borchardt RT. Transport of macromolecules across the capillary endothelium. In: Juliano RL, editor. *Targeted drug delivery*. New York: Springer-Verlag; 1991. p. 43–70.
79. Hawkins RA. Transport of essential nutrients across the blood–brain barrier of individual structures. *Fed Proc* 1986;**45**:2055–9.
80. Kniesel U, Wolburg H. Review. Tight junctions of the blood–brain barrier. *Cell Mol Neurobiol* 2000;**20**:57–76.
81. Wolburg H, Lippoldt A. Review. Tight junctions of the blood–brain barrier: development, composition and regulation. *Vasc Pharmacol* 2002;**38**:323–37.
82. Robakis NK. Beta-amyloid and amyloid precursor protein. Chemistry, molecular biology and neuropathology. In: Terry RD, Katzman R, Bick KL, editors. *Alzheimer disease*. New York: Raven; 1994. p. 317–26.
83. D'Andrea MR, Reiser PA, Polkovitch DA, Branchide B, Hertzog BH, Belkowski S, et al. The use of formic acid to embellish amyloid plaque detection in Alzheimer's disease tissues misguides key observations. *Neurosci Lett* 2003;**342**:114–18.
84. Kawai M, Kalaria RN, Harik SI, Perry G. The relationship of amyloid plaques to cerebral capillaries in Alzheimer's disease. *Am J Pathol* 1990;**137**:1435–46.
85. Thomas T. Beta amyloid mediated vasoactivity and vascular endothelial damage. *Nature* 1996;**380**:168–71.
86. Glenner GG, Wong CW. Alzheimer's disease and Down's syndrome: sharing of a unique cerebrovascular amyloid fibril protein. *Biochem Biophys Res Commun* 1984;**122**:1131–5.
87. Marco S, Skaper SD. Amyloid β-peptide1-42 alters tight junction protein distribution and expression in brain microvessel endothelial cells. *Neurosci Lett* 2006;**401**:219–24.
88. Dede DS, Yavuz B, Yavuz BB, et al. Review. Assessment of endothelial function in Alzheimer's disease: is Alzheimer's disease a vascular disease? *Am Geriatr Soc* 2007;**55**:1613–17.
89. Anfuso CD, Lupo G, Alberghina M. Amyloid β but not bradykinin induces phosphatidylcholine hydrolysis in immortalized rat brain endothelial cells. *Neurosci Lett* 1999;**271**:151–4.
90. Eisenhauer PB, Johnson RJ, Wells JM, Davies TA, Fine RE. Toxicity of various amyloid β peptide species in cultured human blood–brain barrier endothelial cells: increased toxicity of Dutch-type mutant. *J Neurosci Res* 2000;**60**:804–10.
91. Xu J, Chen S, Ku G, Ahmed SH, Chen H, Hsu CY. Amyloid β peptide-induced cerebral endothelial cell death involves mitochondrial dysfunction and caspase activation. *J Cereb Blood Flow Metab* 2001;**21**:702–10.
92. Bruggmann D, Lips KS, Pfeil U, Haberberger RV, Kummer W. Rat arteries contain multiple nicotinic acetylcholine receptor α-subunits. *Life Sci* 2003;**72**:2095–9.
93. Peña VB, Bonini IC, Antollini SS, Kobayashi T, Barrantes FJ. Alpha 7-type acetylcholine receptor localization and its modulation by nicotine and cholesterol in vascular endothelial cells. *J Cell Biochem* 2011;**112**(11):3276–88.
94. Lupo G, Anfuso CD, Assero G, Strosznajder RP, Walski M, Pluta R, et al. Amyloid β(1-42) and its β(25-35) fragment induce in vitro phosphatidylcholine hydrolysis in bovine retina capillary pericytes. *Neurosci Lett* 2001;**303**(3):185–8.
95. Sagare AP, Bell RD, Zhao Z, Ma Q, Winkler EA, Ramanathan A, et al. Pericyte loss influences Alzheimer-like neurodegeneration in mice. *Nat Commun* 2013;**4**:2932.
96. Wilhelmus MM, Otte-Holler I, van Triel JJ, Veerhuis R, Maat-Schieman ML, Bu G, et al. Lipoprotein receptor-related protein-1 mediates amyloid-β-mediated cell death of cerebrovascular cells. *Am J Pathol* 2007;**171**(6):1989–99.

INTRACELLULAR CONSEQUENCES OF AMYLOID IN ALZHEIMER'S DISEASE

97. Kalaria RN, Premkumar DR, Pax AB, Cohen DL, Lieberburg I. Production and increased detection of amyloid β protein and amyloidogenic fragments in brain microvessels, meningeal vessels and choroid plexus in Alzheimer's disease. *Brain Res Mol Brain Res* 1996;**35**:58–68.

98. Blaise R, Mateo V, Rouxel C, Zaccarini F, Glorian M, Bereziat G, et al. Wild-type amyloid β1-40 peptide induces vascular smooth muscle cell death independently from matrix metalloprotease activity. *Aging Cell* 2012;**11**:384–93.

99. D'Andrea MR, Nagele RG. Morphologically distinct types of amyloid plaques point the way to a better understanding of Alzheimer's disease pathogenesis. *Biotech Histochem* 2010;**85**(2):133–47.

100. Clifford PM, Siu G, Kosciuk M, Levin EC, Venkataraman V, D'Andrea MR, et al. Alpha7 nicotinic acetylcholine receptor expression by vascular smooth muscle cells facilitates the deposition of Abeta peptides and promotes cerebrovascular amyloid angiopathy. *Brain Res* 2008;**1234**:158–71.

101. Chi JT, Rodriguez EH, Wang Z, Nuyten DS, Mukherjee S, van de Rijn M, et al. Gene expression programs of human smooth muscle cells: tissue-specific differentiation and prognostic significance in breast cancers. *PLoS Genet* 2007;**3**:1770–84.

102. Cheung C, Goh YT, Wu C, Guccione E. Modeling cerebrovascular pathophysiology in amyloid-β metabolism using neural-crest-derived smooth muscle cells. *Cell Rep* 2014;**9**:391–401.

103. D'Andrea MR. *Bursting neurons and fading memories*. New York: Elsevier Press; 2014.

104. Prior R, Wihl G, Urmoneit B. Apolipoprotein E, smooth muscle cells and the pathogenesis of cerebral amyloid angiopathy: the potential role of impaired cerebrovascular Aβ clearance. *Ann NY Acad Sci* 2000;**903**:1806.

105. D'Andrea MR, Nagele RG. MAP-2 immunolabeling can distinguish diffuse from dense-core amyloid plaques in brains with Alzheimer's disease. *Biotech Histochem* 2002;**77**(2):95–103.

106. Davis-Salinas J, Saporito-Irwin SM, Cotman CW, Van Nostrand WE. Amyloid β-protein induces its own production in cultured degenerating cerebrovascular smooth muscle cells. *J Neurochem* 1995;**65**:931–4.

107. Revesz T, Holton JL, Lashley T, Plant G, Rostagno A, Ghiso J, et al. Sporadic and familial cerebral amyloid angiopathies. *Brain Pathol* 2002;**12**:343–57.

108. Qiu Z, Strickland DK, Hyman BT, Rebeck GW. Alpha2-macroglobulin enhances the clearance of endogenous soluble β-amyloid peptide via low-density lipoprotein receptor-related protein in cortical neurons. *J Neurochem* 1999;**73**:1393–8.

109. Kanekiyo T, Zhang J, Liu Q, Liu CC, Zhang L, Bu G. Heparan sulphate proteoglycan and the low-density lipoprotein receptor-related protein 1 constitute major pathways for neuronal amyloid-β uptake. *J Neurosci* 2011;**31**:1644–51.

110. Wyss-Coray T, Loike JD, Brionne TC, Lu E, Anankov R, Yan F, et al. Adult mouse astrocytes degrade amyloid-β in vitro and in situ. *Nat Med* 2003;**9**:453–7.

111. Kanekiyo T, Liu C-C, Shinohara M, Li J, Bu G. LRP1 in brain vascular smooth muscle cells mediates local clearance of Alzheimer's amyloid-β. *J Neurosci* 2012;**32**(46):16458–65.

112. Nixon RA, Mathews PM, Cataldo AM. The neuronal endosomal–lysosomal system in Alzheimer's disease. *J Alzheimers Dis* 2001;**3**:97–107.

113. Hu X, Crick SL, Bu G, Frieden C, Pappu RV, Lee JM. Amyloid seeds formed by cellular uptake, concentration, and aggregation of the amyloid-β peptide. *PNAS* 2009;**106**:20324–9.

Implications
of Intraneuronal Aβ

*A*lzheimer's disease is not reversible; if you have it, you cannot cure it because once the neurons die, they cannot be resurrected.

Therefore, the only course is prevention. Prevention can only occur before the diagnosis of AD. Prognostic biomarkers such as vascular risk factors, Aβ42/Aβ40 plasma concentration ratios, etc. are required to treat those that have the highest risk of developing AD in an effort to treat the disease at earlier stages. And, there must be a comprehensive understanding of how the disease begins and continues to deteriorate in order to intervene, which may include the redefining of the therapeutic target(s).

Intracellular Consequences of Amyloid in Alzheimer's Disease.
DOI: http://dx.doi.org/10.1016/B978-0-12-804256-4.00009-7

WHY HAVE THE CLINICAL TRIALS FAILED?

I present a series of quotes from some of the authors cited in this book to address this question concerning the ineffectiveness of the amyloid hypothesis-based clinical trials. I find their messages more compelling than summary statements.

- "The amyloid hypothesis has driven drug development strategies for AD for over 20 years." "The amyloid hypothesis has become difficult to challenge because it is so often the lens through which peer reviewers, granting bodies and pharmaceutical companies view, judge and support AD research. Thus new non-amyloid data tends to be couched in terms that place it within the amyloid hypothesis and many authors tacitly ignore valid, but quite different, interpretations."[1]
- "Many years of intense research and millions of research dollars did not help to resolve the role for Aβ in AD. Until recently, the amyloid hypothesis was the major hypothesis of AD research, however, these scientific facts make it hard to accept the amyloid cascade hypothesis. Over the past decade, the amyloid cascade hypothesis has gradually become a dogma. Yet, the amyloid hypothesis has not successfully attempted to explain the disease pathogenesis through the prism of pathogenetic primacy of amyloid deposition in the brain tissue of affected individuals."[2]
- "Based on *in vitro* and *in vivo* electrophysiological and behavioral studies of Aβ protein, it is impossible to conclude their relevance to brain physiology or AD." [The data from transgenic model of AD imply] "a need to critically re-evaluate another dogma, a universal belief that transgenic animals expressing normal or mutated transgene for Aβ represent a true model for AD type neurodegeneration and also brings into doubt the validity of the amyloid cascade hypothesis."[2]
- "A central question is whether Aβ plays a direct role in the neurodegenerative process in AD. There are two schools of thought on involvement of Aβ in AD. In one, Aβ initiates the disease once produced in excess, which has motivated most AD clinical trials. Alternatively, Aβ does not initiate but rather is secondary to other pathogenic events as a protective response to neuronal insult."[3]
- "Abnormalities in synaptic function resulting from the absence or inhibition of the Aβ-producing enzymes suggest that Aβ itself may have normal physiological functions which are disrupted by abnormal accumulation of Aβ during AD pathology. This interpretation suggests that AD therapeutics targeting the β- and γ-secretases should be developed to restore normal levels of Aβ or

combined with measures to circumvent the associated synaptic dysfunction(s) in order to have minimal impact on normal synaptic function."[4]

- "Accumulating data on the biological roles of BACE1, particularly evidence that completes inhibition of BACE1 activity which is deleterious for normal neuronal function, suggests caution for using BACE1 inhibitors as a treatment for AD. In order to improve the development of effective therapeutics that target this enzyme, there is a need to identify ways to avoid the synaptic dysfunction associated with blocking BACE1, which may include partial inhibition strategies." "Many BACE1 inhibitors have been shown to decrease soluble Aβ production, amyloid plaque deposition, as well as improve cognitive function in AD animal models. Surprisingly, none of them have been tested to determine their ability to improve synaptic dysfunction, the cellular mechanism that correlates with cognitive decline. A critical question is whether these inhibitors can recover synaptic deficits seen in AD models or whether they may produce additional defects as seen in BACE1 -/- mice."[4]
- "A therapeutic approach based on Aβ vaccination can be also detrimental and not exclusively protective, as previously suggested in transgenic mouse models of AD. The data challenges the hypothetical idea to use Aβ vaccination as a treatment option for AD."[5]

BUM STEER

The whole concept of deposition has clouded judgment. When intraneuronal Aβ was first visualized many years ago, it was merely considered the primary source of the extracellular deposition, thereby setting the focus of the pathology outside the neurons. This single paradigm is the basis of thousands of publications, millions of dollars, and several clinical trials. Because of amyloid deposition, the production and processing of neuronal Aβ has been well studied, the mechanisms of plaque development (from diffuse to senile, dense-core with the assistance of the inflammatory cells) have been determined, and Aβ's toxic fragments characterized. This bias made the research path so deeply etched that any deviations would be considered outliers.

The lack of a plaque nomenclature has led to misinformation. Amyloid plaques are not associated with cognitive decline because they are observed in normal, nondemented patients, and because they are not present before the onset of symptoms of cognitive decline. These observations can be explained. Not all plaques have the same origin.[6,7] The diffuse amyloid plaques are pools of extracellular Aβ from leaks in the vascular system, are not associated with inflammatory cells, and are not associated with cognitive decline, which may have been identified in those studies. In contrast, the dense-core, senile plaques arise from the lysis of the neurons and associated with cognitive decline, and neuroinflammation. As the neurons continue to accumulate Aβ, they begin to degenerate resulting in synaptic collapse, abnormal signaling, and cognitive decline; all before the plaque (or dead neuron) appears, thereby explaining the enigma. This process also occurs in Aβ-overburdened, vascular smooth muscle cells, endothelial cells, and pericytes. This plaque-type distinction is critical when reporting any information associating plaques with any clinical or neuropathological feature.[6,7]

Of the hundreds of papers and editorials acknowledging the failure of the amyloid-deposition hypothesis and of the clinical trials, many point out flaws in the target, but rarely has anyone presented a new model that can accommodate most of the salient pathological reports presented in this book.

THE LYTIC MODEL

The name of this model was revised from "Inside-Out," my initial name to describe the process by which neurons burst from the excessive accumulation of Aβ "inside" to release their neuronal contents "out" into the parenchyma as presented in a short YouTube video.[8] When I published this body of work in my previous book entitled *Bursting Neurons and Fading Memories*, one of the reviewers referred to it as the "Bursting" hypothesis, but through writing this book and in the context of the other amyloid models that are based on deposition, the most distinguishing name is the Lytic Model (see Chapter 5).[9]

Lysis (or lytic) immediately denotes cell death, which is the foundation of this model. Not only neurons, but smooth muscle and endothelial cells, and pericytes, die from overaccumulation of Aβ independent of deposition (a critically distinguishing postulate in this model). This model does not just describe how some neurons are dying, but it also represents the entire sequence of pathological events leading to AD. The model begins at the vascular level and ends with the spiraling out-of-control cycles of neuronal death, which also includes secondary neuronal death independent of intracellular Aβ by (1) inflammatory mediators and (2) damaged or digested processes within the lytic zone of nearby, and perhaps distant, neurons.

The Lytic Model is characterized by several time-lapsed phases that are orchestrated by four essential constituents: the vascular system, the neurons, the inflammatory cells, and Aβ. The description of each phase is accompanied with an illustration and supportive images. I also provide supportive citations composed of quoted text from authors cited in this book for similar reasons noted above. I find this format most effective and convincing to allow the readers to read original text without any biased rendition.

Phase 0: Controls

Normal physiological levels of Aβ are present in neurons and other cells; the blood−brain barrier (BBB) is functional; the patient has no signs of vascular pathological or measurable cognitive impairment. However, the presence of vascular risk factors will move the patient from Phase 0 to Phase 1 (Fig. 9.1) (see Chapter 8). This phase accounts for all the natural functions of Aβ and its receptors in neuronal viability, learning and memory, synaptic plasticity, wound healing, etc., which

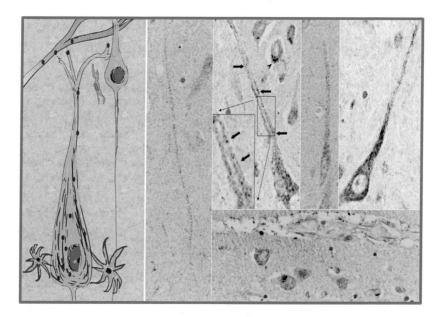

FIGURE 9.1   Control brain (Phase 0). Illustration shows two neurons (*blue*) with natural levels of intraneuronal Aβ (*brown granules*), two neuronal-associated astrocytes (*red*), blood vessel (top) with intravascular Aβ42 (*brown granules*), and an unreactive, resting microglial cell (*pink*). No vascular leaks are observed. Natural levels of intraneuronal Aβ42 are present in nondemented human brain neurons (IHC panels on the right). *Aβ42*, amyloid-β42; *IHC*, immunohistochemistry. *Source: Used with permission from Slidomics, LLC;* Biotech Histochem *2010;85(5):133−47.*

also explains why inhibiting Aβ will lead to deleterious effects as also noted in the first section of this chapter (see Chapter 3).

Supporting Citations

- "Inhibition of endogenous Aβ production by exposure to inhibitors either of β- or γ-secretases in primary neuronal cultures caused neuronal cell death. Targeting Aβ formation pharmacologically, or immunologically, could be deleterious. Both Aβ40 and Aβ42 have been shown to be protective. Aβ may have an important physiological role in synapse elimination during brain development."[9]
- "Given the important role that Aβ plays in various activities at the synapse, Aβ should not be regarded merely as a toxic factor that requires eradication to avoid dementia. There is enough evidence to suggest an essential activity-dependent role of Aβ in modulation of synaptic activity and neuronal survival."[9]
- "The results established BACE1 and APP processing pathways as critical for cognitive, emotional, and synaptic functions, and future studies should be alert to potential mechanism-based side effects that may occur with BACE1 inhibitors designed to ameliorate Aβ amyloidosis in AD."[10]
- "By 6 months, PS cDKO [presenilin 1 and 2 double knockout mice] mice showed even greater synaptic deficits, including loss of presynaptic inputs and enhanced basal synaptic transmission, in addition to reduced LTP and PPF [paired-pulse facilitation] ratio. These synaptic impairments may explain the age-dependent deterioration in the cognition of the PS cDKO mice. Collectively, these studies suggest that presenilins are essential for synaptic plasticity as well as learning and memory in the adult brain."[4]

Phase 1: Vascular Pathology

Patients with high cardiovascular risks eventually develop vascular lesions, including endothelial cell degeneration, which over time leads to focal areas of intracellular Aβ in other local vascular cells including the smooth muscle cells and pericytes that degenerate over time leading to BBB dysfunction (see Chapter 8). Note that similar issues are expected to happen at the blood−retinal barrier.[7] These vascular pathological events lead to cell death and the formation of vascular-associated dense plaques that may trigger an inflammatory response. The dysfunctional BBB leaks unregulated blood components including Aβ fragments and immunoglobulins into the brain. Aβ pools in local areas forming diffuse, vascular-associated plaques, which do not trigger an inflammatory response. Continual vascular leakage leads to

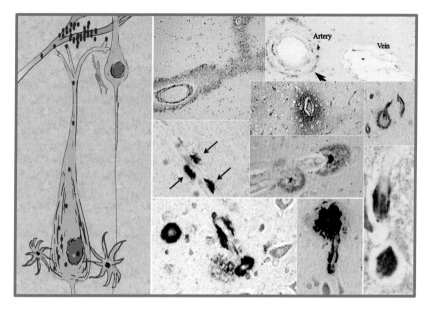

FIGURE 9.2 Vascular pathology (Phase 1). Illustration shows two neurons (*blue*) with natural levels of intraneuronal Aβ (*brown granules*), two neuronal-associated astrocytes (*red*), and blood vessel (top) with intravascular Aβ42 (*brown granules*) that is leaking out due to the dysfunctional BBB. Leaked Aβ42 pools to form diffuse plaques, also present is an unreactive, resting microglial cell (*pink*). IHC panels on the right show vascular-associated diffuse plaques (top left), other panels show dense formations of vascular-Aβ42 plaques believed to be pericytes, and smooth muscle cells. One image (middle center below the artery/vein panel) shows the leakage of human immunoglobulins (*brown stained*) out of a vessel into the brain. *Aβ42*, amyloid-β42; *BBB*, blood−brain barrier; *IHC*, immunohistochemistry. *Source: Used with permission from Slidomics, LLC; Biotech Histochem 2010;85(5):133−47.*

lower concentrations of plasma Aβ42 (Fig. 9.2). This phase accounts for the list of cardiovascular benefits (eg, exercise, mental gaming) to help keep the brain perfused while sustaining a healthy BBB.[11,12]

Supporting Citations

- "BBB breakdown, due to disruption of the tight junctions, altered transport of molecules between blood and brain and brain and blood, aberrant angiogenesis, vessel regression, brain hypoperfusion, and inflammatory responses, may initiate and/or contribute to a 'vicious circle' of the disease process, resulting in progressive synaptic and neuronal dysfunction and loss in disorders such as Alzheimer's disease."[13]
- "Earlier studies in [amylotrophic lateral sclerosis] patients have suggested a possible blood [cerebral spinal fluid] barrier and BBB

breakdown by demonstrating increased levels of albumin, [immunoglobulin G], and complement components in the CSF or in the spinal cord."[13]

- "Vaccination with amyloid-β peptide induces autoimmune encephalomyelitis in C57/BL6 mice," and may explain why 15 patients (or 6%) showed clinical signs consistent with CNS inflammation in a suspended clinical trial. "The experimental disease was observed [in mice] only when pertussis toxin . . . was coadministered."[5] For the monoclonal antibody to gain entry into the brain of the patients with mild to moderate AD patients, the BBB must be compromised, which is replicated in the mouse model only with the use of pertussis toxin that is derived from *Bordetella pertussis* and is known to increase the permeability of the BBB.[14]

- "Since the value of scientific evidence generally revolves around probability and chance, it is concluded that the data presented here poses a powerful argument in support of the proposal that AD should be classified as a vascular disorder. According to elementary statistics, the probability or chance that all these findings are due to an indirect pathological effect or to coincidental circumstances related to the disease process of AD seems highly unlikely. The collective data presented in this review strongly support the concept that sporadic AD is a vascular disorder."[15]

- "The adjusted risks for dementia were significantly inversely associated with increased total or daily equivalent statin dosage. Patients who received the highest total equivalent doses of statins had a fold decrease in the risk of developing dementia. Similar results were found with the daily equivalent statin dosage."[16]

- "In addition to simply coexisting together, VaD [vascular dementia] and AD share certain vascular risk factors such as hypertension, peripheral arterial disease, some cardiovascular disorders, diabetes mellitus, and smoking that can complicate the diagnostic process. These mutual risk factors tend to indicate that similar mechanisms may be involved in the pathogenesis of both disorders, including breach of the blood-brain barrier, APOE, oxidative stress, renin-angiotensin system derangements, apoptosis, neurotransmitter abnormalities, and psychological stress."[17]

- "A prospective population-based cohort study found that higher serum uric acid levels are related to a decreased risk of dementia and better cognitive function later in life, but only after adjustment for several cardiovascular risk factors. This corroborates the notion that oxidative stress is involved in the pathogenesis of dementia and cognitive impairment and suggests a possible protective role for antioxidants such as uric acid."[18]

- "Retinal blood flow in MCI is intermediate between what is measured in control subjects and in AD patients. Our findings suggest that blood flow abnormalities may precede the neurodegeneration in AD."[19]
- "A breakdown of BBB was evident in young 4- to 10-month-old Tg2576 mice. Compromised barrier function could explain the mechanisms of Aβ entry into the brain observed in experimental Alzheimer disease vaccination models. Such structural changes to the BBB caused by elevated Aβ could play a central role in Alzheimer disease development and might define an early point of intervention for designing effective therapy against the disease."[20]
- "Long-term aspirin therapy brings about a 20%−25% reduction in the odds of subsequent myocardial infarction, stroke, or vascular death among intermediate- or high-risk cardiovascular disease patients. Recent studies have shown that it is also an effective treatment for AD patients. In human studies, users of high-dose aspirin had significantly lower prevalence of AD and better-maintained cognitive function than nonusers. However, aspirin needs to be taken before the symptoms of AD occur. It had no effect on AD at a later stage when the brain damage is severe. These results suggest that repurposing existing antiplatelet drugs for the treatment of AD may be beneficial,"[21] and speak to the irreversibility of the disease.

Phase 2: Neuronal Overaccumulation

In addition to the overaccumulation of intracellular Aβ in the vascular cells, unregulated Aβ continues to seep into the brain through the dysfunctional BBB, and binds to and is endocytosed by the Aβ-high-affinity neuronal receptors (Fig. 9.3). This phase accounts for the benefits of nicotine to compete against Aβ for the α7 receptors in AD trials thereby reducing toxic amounts of intraneuronal Aβ. The presence of cerebral amyloid angiopathy (CAA), synaptic degeneration, and initial cognitive impairment occurs during this phase and before neuronal death (see Chapters 6, 8). This phase also accounts for the lack of efficacy in removing extracellular Aβ since this diffuse-plaque amyloid is not associated with neuronal death. Also note that there continues to be no evidence of "extracellular" Aβ42 accumulating on or nearby the neurons as per the amyloid-deposition hypotheses.

Supporting Citations

- "The combination of pharmacological, biochemical, and fluorescence microscopy tools showed that α7AChRs [nicotinic acetylcholine

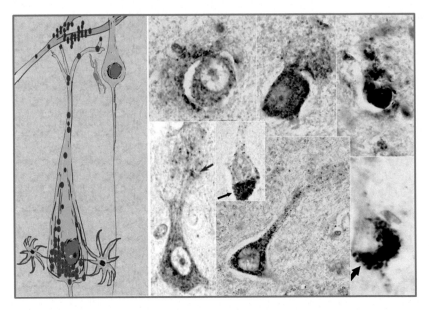

FIGURE 9.3   Neurons overaccumulate Aβ42 (Phase 2). Illustration shows two neurons
(*blue*) with higher levels of intraneuronal Aβ mediated by receptors (*red-outlined, brown
granules*), two neuronal-associated astrocytes (*red*), and blood vessel (top) with intravascular Aβ42 (*brown granules*) that is leaking out due to the dysfunctional BBB. IHC panels
show various levels of intraneuronal overaccumulation to the point when the neuron is
about ready to lyse. At this time, synaptic failure has begun, and other intraneuronal
processes are crippled. *Aβ42*, amyloid-β42; *BBB*, blood–brain barrier; *IHC*, immunohistochemistry. *Source: Used with permission from Slidomics, LLC;* Biotech Histochem
2010;85(5):133–47.

receptors] in rat arterial endothelial and human venous endothelial
cells occur at extremely low expression levels (∼50 fmol/mg protein)
but undergo agonist-induced up-regulation at relatively high
nicotine concentrations (∼300-fold with 50 microM ligand),
increasing their cell-surface exposure. When analysed in terms of
cold Triton X-100 solubility and subcellular distribution, α7AChRs
occur in the 'non-raft' subcellular membrane fractions,"[22] which
could explain the overaccumulation of intracellular Aβ in virtually
any cell type with these receptors including the vascular cells and
neurons (see Chapter 8).
- "Chronic exposure of these cells [*Xenopus* oocytes, fibroblast cell line]
  to nicotine or another agonist is shown to result in an increase in
  receptor amount, indicating that nicotine-induced up-regulation
  reflects properties of the α4β2 receptor protein, rather than being an
  adaptive response unique to the neurons in which these receptors are
  normally expressed. Up-regulation does not appear to require ion

flow through the ion channel, because it is also caused by
mecamylamine, which blocks the ion channel, and because after
prolonged exposure to nicotine most receptors become permanently
unable to open their channels in response to nicotine binding. The
noncompetitive antagonist mecamylamine blocks open channels
more effectively, and so it is more effective at blocking channels in
the presence of nicotine. Mecamylamine and nicotine are also
synergistic in causing receptor up-regulation. Ligands that cause up-
regulation appear to induce a conformation of the receptor that is
removed from the surface and degraded more slowly,"[23] thereby
providing a receptor-mediated mechanism for the overaccumulation
of intracellular Aβ in neurons and vascular cells.

- "Overall, there is accumulating evidence to suggest that
intraneuronal Aβ42 is a major risk factor for neuron loss and a trigger
for the Aβ cascade of pathological events."[24]
- "In conclusion, multiple, nonmutually exclusive, molecular
mechanisms have been proposed to account for the neuroprotective
effects of nicotine. These mechanisms range from direct or indirect
blockade of the toxic agent itself, to induction of neurotrophic agents
or activation of intracellular antiapoptotic cascades in targeted cells.
These nAChR [nicotinic acetylcholine receptors]-related protective
events may be, at least in part, triggered by calcium entry or
activation of enzymes associated with the receptor itself."[25]
- "The unusual nature of Aβ42-induced receptor activation and
desensitization indicates that the *in situ* effects on α7nAChRs could
prove to be quite complex. In general, however, the data indicated
that compounds targeted to blocking the effects of Aβ on α7nAChR
function may be a promising therapy for AD."[26]
- "The study of the expression of specific nicotinic receptors in
different cell types and organs, and the development of selective
nicotinic agonists is critical to develop novel pharmacological
strategies against neurological, infectious and inflammatory
disorders."[27]
- "In a triple-transgenic AD mouse model with the PS1 (M146V), APP
(Swe), and tau (P301L) transgenes, deficits in synaptic transmission
and long-term potentiation were observed before plaque and tau
pathology, but in the presence of early intraneuronal Aβ
accumulation. It has been suggested that intraneuronal Aβ
pathology causes the onset of early cognitive deficits in this triple-
transgenic AD mouse model. In other experiments, it has been
shown that intracellular Aβ accumulation directly affects cell
viability, ultimately resulting in neuronal death. These observations
raise the hypothesis that intraneuronal Aβ accumulation may be one
of the initial steps in a cascade of events leading to AD."

"Furthermore, reducing the endocytosis of Aβ by neurons may have a therapeutic effect on AD."[28]

- "In summary, the results presented in this paper demonstrate that the α7nAChR knockout mutation rescues a significant part of the human APP over-expressing phenotype. Thus, the α7nAChR contributes to both the pathology and behavioral deficits observed in the APP over-expressing mouse. These results are consistent with the hypothesis that the α7nAChR plays a role in AD and suggests that blocking the α7nAChR could alleviate some symptoms of AD."[29]
- "Taken together, our findings provide novel insights into the molecular mechanisms of AD and CAA, and suggest that restoring LRP1 expression in the brain vasculature in patients with AD and CAA could be explored for therapy. It might be possible to screen for chemical compounds that can increase LRP1 expression and/or function in vascular smooth muscle cells. Several compounds including rifampicin and caffeine seem to upregulate LRP1 expression and accelerate Aβ clearance in cerebrovasculature in mice. These compounds can be explored as novel therapies for both AD and CAA."[30]
- "In view of the pathophysiological consequences of Aβ42 binding to α7nAChR on neuronal surfaces that stem from excessive intraneuronal Aβ42 accumulation, the α7nAChR could be an important therapeutic target for treatment of AD. In addition, it further emphasizes the potential merits of new and effective therapeutic strategies pointed towards the goal of lowering of Aβ42 levels in the blood and cerebrospinal fluid as well as blocking Aβ42 in the blood from penetrating the blood-brain barrier and entering into the brain parenchyma."[31]
- "It can be predicted that blockade of exogenous Aβ42 from entering vulnerable pyramidal neurons *in vivo* will reduce intracellular Aβ42 accumulation. This will in turn, result in prolonging neuron survival time and hence, slow down the degeneration process."[32]

Phase 3: Neuronal Lysis (Plaque Formation)

Excessive accumulation of intraneuronal Aβ eventually leads to neuronal death as senile, dense-core plaques (Fig. 9.4). Proteolytic-resistant neuronal debris remains in place and can be easily detected (eg, ubiquitin, cathepsin D, tau, and DNA). The released neuronal-lysosomal enzymes digest all proteolytic-sensitive proteins (eg, MAP2 in Fig. 9.5) radially in a fairly defined area, thereby producing the typical spherically shaped plaque. This phase also accounts for the lack of efficacy in removing extracellular Aβ plaques since the neurons have already died.

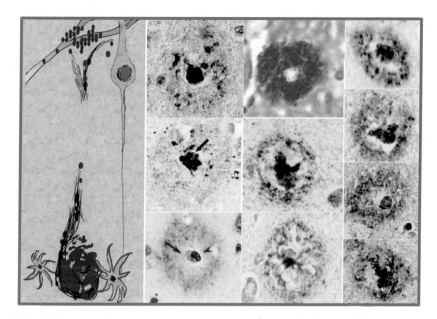

FIGURE 9.4   Plaque formation (Phase 3). Illustration shows the lysed neuron that dies in its place to form the senile, dense-cored plaque. The resting microglial may react to the synaptic debris at this time. IHC panels show the dense-core, senile plaques, most still retain their hematoxylin, blue-stained nucleus. Bottom 2 panels (left and center) show the residual neuronal lipofuscin in place (yellow pigment in left bottom panel [*large arrow*], and red-stained material in center bottom panel). Top middle shows the missing DNA after this tissue was first pretreated with DNAse before the IHC stain, and the third panel from the top on the right shows the presence to red-labeled, NeuN in the dead-center of a plaque. All plaques show the characteristic radial diffusion of the Aβ42 (red-, brown-, or black-stained). *Aβ42*, amyloid-β42; *BBB*, blood−brain barrier; *IHC*, immunohistochemistry; *NeuN*, neuronal-specific nuclear protein. *Source: Used with permission from Slidomics, LLC;* Biotech Histochem *2010;85(5):133−47.*

Supporting Citations

- "The staining of a number of enlarged neurites is associated with senile plaques and a high density of neurites not obviously associated with senile plaques or NFT. Similar observations were seen with six other Alzheimer patients ranging in age at death from 69 to 82 years. Preliminary studies indicated that these three types of ubiquitin-positive profiles are most abundant in regions where senile plaques and NFT are abundant, as defined by Congo red staining."[33]
- "The presence of repetitive centromeric DNA sequences dispersed throughout a dense-core plaque lacking a detectable neuronal nuclear remnant. We consider these centromeric sequences to represent residual DNA fragments that are eventually dispersed throughout the plaque after degradation of the neuronal nuclei."[6]

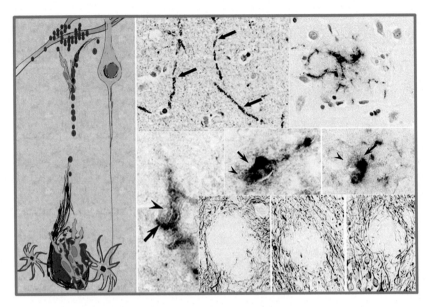

FIGURE 9.5   Neuroinflammation begins with the reactive microglia movement to the lysed neurons (Phase 4a). Illustration shows the lysed neuron that dies in its place to form the senile, dense-cored plaque that now contains reactive microglia. IHC panels show thick, dystrophic, Aβ42-ladened neuronal processes (second from top left panel). Reactive microglia are detected in the center of these now toxic plaques (top panels, *blue* = microglia; *red* = Aβ42), reactive microglia detected in a single IHC stain (second to top right panel). Panels in the top bottom half show the very close association of reactive microglia (*blue*) with NeuN (*red*). The bottom panels show the destructive of the lysed neuron that digested MAP2-immunoreactivity. *Aβ42,* amyloid-β42; *IHC,* immunohistochemistry; *MAP2,* microtubule-associated protein 2; *NeuN,* neuronal-specific nuclear protein. *Source: Used with permission from Slidomics, LLC;* Biotech Histochem *2010;85(5):133–47.*

- "[Emil] Redlich observed the presence of large numbers of plaques in grey matter, often in proximity to degenerating ganglion cells [in 1898]. They were of different sizes, contained some fibrous material and tended to have a nucleus in the centre. He referred to the smaller plaques as cotton wool-like, introducing yet another term that is still in use."[34]
- "Results show that both Aβ42- and cathepsin D-immunoreactivity are found throughout the volume of the same amyloid plaque and in similar locations within neurons. Since cathepsin D is highly localized to the lysosomal compartment, it is concluded that the Aβ42-positive granules that dominate the cytoplasmic volume within the neurons of AD brains most likely represent lysosomes and/or their derivatives."[35]
- "Neurofibrillary tangles, tau, neurofilament, and ubiquitin in plaques, and in nearby neurons."[6]

- "Predominance of neuronal mRNAs in individual Alzheimer's disease senile plaques. The mRNAs in SPs [senile plaques] were compared with those in individual CA1 neurons and the surrounding neuropil of control subjects. The remarkable demonstration here, that neuronal mRNAs predominate in SPs, implies that these mRNAs are nonproteinaceous components of SPs, and, moreover, that mRNAs may interact with Aβ protein and that SPs form at sites where neurons degenerate in the AD brain."[36]
- "The fact that Aβ42 immunolabeling was largely restricted to neurons and amyloid plaques supports a neuronal origin for dense-core amyloid plaques. In amyloid plaques, lipofuscin tends to remain centrally located."[37]
- "This is the first report of the presence of NeuN [a neuronal-specific nuclear protein] in the dense-core region of amyloid plaques in AD brains. The pathological mechanisms leading to the presence of a NeuN-positive neuronal nucleus in these plaques are unknown, but this observation is consistent with the previous investigation that showed that these plaques are stained similarly with DAPI [a DNA-specific fluorescent nuclear stain]. The presence of NeuN-positive neuronal nuclear remnants within plaques further supports a neuronal origin of dense-core plaques. These neuron-derived nuclear remnants are readily distinguished from the nuclei of microglia and astrocytes by virtue of the fact that latter invariably are much smaller."[6]

Phase 4a: Neuroinflammation (Microglia Discovery)

Reactive microglia migrate to the released DNA/ATP deep within the lysed neurons (Fig. 9.5). This phase accounts for the presence of microglia in the center of these dense-core plaques in close association with the neuronal nuclei. The reactive microglia also seem disinterested in the Aβ42, which in stark contrast to the report that microglia react to extracellular Aβ to form (or mature) plaques. This phase accounts for the limited success of NSAIDs and other anti-inflammatory therapeutics because even though the neuron has died (primary event), secondary neuronal cell death due to inflammatory mediators may be spared (see Chapter 4).

Supporting Citations

- "Additional evidence from transgenic mice and from non-demented patients with amyloid plaques suggests that fibrillar Aβ alone may not be sufficient to initiate brain inflammation," leaving neuronal death as the true initiator of neuroinflammation.[38]
- Long-term NSAID usage effectively and significantly reduced the numbers of microglia (~threefold) without affecting the number of

diffuse or neuritic (dense-core) plaques in nondemented patients as compared with similar patients with no history of arthritis or other condition that might promote regular NSAIDs usage. "These results suggest that if NSAID use is effective in treating AD, the mechanism is more likely to be through the suppression of microglial activity than by inhibiting the formation of SP or neurofibrillary tangles,"[39] which (1) is contrary to the widely accepted theory of plaque formation through the processed of neuroinflammation and (2) supports the neuronal Lytic Model of senile plaque formation.

- "Differential expression of each isoform may play a regulatory role in the physiological and pathophysiological functions of RAGE that binds Aβ. Analysis of RAGE expression in non-demented and AD brains indicated that increases in RAGE protein and percentage of RAGE-expressing microglia paralleled the severity of disease."[40]
- "At the ultrastructural level, cathepsin immunoreactivity in senile plaques was localized principally to lysosomal dense bodies and lipofuscin granules, which were extracellular. Similar structures were abundant in degenerating neurons of Alzheimer neocortex, and cathepsin-laden neuronal perikarya in various stages of disintegration could be seen within some senile plaques. The high levels of enzymatically competent lysosomal proteases abnormally localized in senile plaques represent evidence for candidate enzymes that may mediate the proteolytic formation of amyloid."[41]
- "[It was] reported that extracellular ATP induces membrane ruffling and chemotaxis of microglia and suggested that their induction is mediated by the Gi/o-protein-coupled P2Y(12) receptor. Here we report discovering that the P2X(4) receptor is also involved in ATP-induced microglial chemotaxis."[42]
- "Extracellular ATP serves as a danger signal to alert the immune system of tissue damage by acting on P2X or P2Y receptors."[43]
- "In brains with damaged neurons and astrocytes, large amounts of nucleotides are released from these cells. The extracellular nucleotides, which are easily diffused and rapidly catalyzed by ATPases, may play a role in modulating the microglial function of the brain in the early phase of pathology. A novel effect was presented of extracellular ATP and adenosine triphosphate (ADP) nucleotides on microglia, that is, the induction of chemotaxis via Gi /o-coupled P2Y receptors."[44]

Phase 4b: Neuroinflammation (Glial Scar Formation) Leading to Collateral Damage

The reactive microglia secrete factors that activate astrocytes to form the glial scar. Meanwhile, the secreted inflammatory factors (cytokines and chemokines) lead to additional neuronal death in the vicinity independent of the levels of Aβ intraneuronal accumulation. Depending on the extent of damage to the collateral processes of the nearby (and perhaps distant) neurons in the toxic area of the lysed neuron, additional neuronal death could occur, while triggering NFT formation in remote neurons (Fig. 9.6). This phase accounts for the formation of the glial scars around this type of plaque and explains the circumstances of additional neuronal death independent of intraneuronal Aβ. This phase may also explain the partial benefits of NSAIDs to reduce the processes of additional neuronal death due to the released inflammatory factors (see Chapter 7).

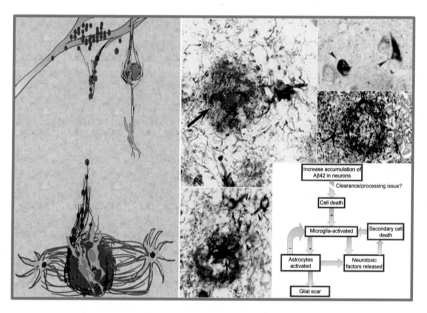

FIGURE 9.6   Neuroinflammation continues with glial scar formation leading to collateral damage (Phase 4b). Illustration shows the lysed neuron that dies in its place to form the senile, dense-cored plaque that now contains reactive microglia, and how the local astrocytes become activated to extend their processes into the neuronal-vacant area. Triple IHC panels reactive microglia (*blue*), and activated astrocytic-GFAP fibers (*black*) in Aβ42 (*red*) plaques. Low panel shows the proposed inflammatory cascade of events based on the literature reporting the effects of secreted factors from the microglia and astrocytes that ultimately end up with the glial scar, but also collateral neuronal deaths, perhaps by NFT formation (top right panel; NFTs: *black*; Aβ42: *red*). A single IHC panel (second from top right) shows the numerous GFAP-positive astrocytic fibers forming the prominent glial scar. *Aβ42*, amyloid-β42; *BBB*, blood–brain barrier; *IHC*, immunohistochemistry; *GFAP*, glial fibrillary acidic protein; *NFT*, neurofibrillary fibrillary tangles. *Source: Used with permission from Slidomics, LLC;* Biotech Histochem *2001;**76**(2);97–106;* Neurobiol Aging *2004;**25**:675–83.*

Supporting Citations

- "As peripheral biomarkers of Alzheimer's disease in the cerebrospinal fluid, Tau proteins are now validated for diagnosis and predictive purposes. The detailed characterization of Tau in the brain and in peripheral fluids will lead to novel promising biomarkers for differential diagnosis of dementia and monitoring of therapeutics."[45]
- "Tangle and neuron numbers, but not amyloid load, predict cognitive status in Alzheimer's disease."[46]
- "(1) The severity of dementia was positively related to the number of NFTs in neocortex, but not to the degree of SP [senile plaque] deposition; (2) NFTs accumulate in a consistent pattern reflecting hierarchic vulnerability of individual cytoarchitectural fields; (3) NFTs appeared in the entorhinal cortex, CA1/subiculum field of the hippocampal formation, and the amygdala early in the disease process; and (4) the degree of SP deposition was also related to a hierarchic vulnerability of certain brain areas to accumulate SPs, but the pattern of SP distribution was different from that of NFT."[47]
- "We observed a consistent and robust induction of NFT formation (e.g. CA1) in regions with only very scarce plaques, but functionally connected to brain regions with high plaque density (subiculum)."[48]

FINAL COMMENTS

The majority of the about 1000 publications cited in this book begin with an introduction of AD, a description of the amyloid-deposition hypothesis, and if the publication is recent, they question the amyloid hypothesis due to failed clinical trials. Others note that amyloid plaques are formed by the deposition of amyloid, and many continue to state that plaques do not correlate to the severity of the disease.

First, I would update such passages to say that "contrary to what has been taught, there is no evidence to prove that amyloid plaques form from extracellular deposition of amyloid secreted from neurons." It is vital to differentiate plaque types since not all amyloid plaques define neuronal death.[6] However, neuronal death defines cognitive impairment. Published illustrations depicting the growth of plaques outside neurons are highly inaccurate.

Finally, the Lytic Model can accommodate many of the experimental findings presented in this book, including intracellular Aβ, neuroinflammation, plaques, vascular pathologies, and risk factors, and can explain the failure of the amyloid-based clinical trials. This model also can explain the genetic forms of AD (eg, FAD, Down syndrome) as an accelerated pathological process. Based on my extensive research, no one has presented such a comprehensive and accommodating model that can claim such a wide range of inclusions. This model renders the vascular-derived Aβ as perhaps one of the most dangerous components to the CNS in the human body.

All signs point to reducing or blocking vascular-derived Aβ (and other components such as immunoglobulins) from entering the brain; hence, a healthy BBB is the first level of defense to prevent AD, and perhaps other CNS degenerative diseases.

Based on the natural requirement for Aβ in many biological processes (see Chapter 3), perhaps it is the **"toxic accumulation" of Aβ in cells** that leads to their death, rather than the accumulation of "toxic Aβ." Therefore, the second level of defense would be to inhibit the overaccumulation of Aβ in cells, especially neurons.

References

1. Morris GP, Clark IA, Visse B. Review. Inconsistencies and controversies surrounding the amyloid hypothesis of Alzheimer's disease. *Acta Neuropathol Comm* 2014;**2**:135.
2. Koudinov AR, Berezov TT. Review. Alzheimer's amyloid-β (Aβ) is an essential synaptic protein, not neurotoxic junk. *Acta Neurobiol Exp* 2004;**64**:71−9.
3. Parihar MS, Brewer CJ. Review. Amyloid beta as a modulator of synaptic plasticity. *J Alz Dis* 2010;**22**(3):741−63.
4. Wang H, Megill A, He K, Kirkwood A, Lee H-K. Consequences of inhibiting amyloid precursor protein processing enzymes on synaptic function and plasticity. *Neural Plasticity* 2012;**272374**:24 pp.
5. Furlan R, Brambilla E, Sanvito F, Roccatagliata L, Olivieri S, Bergami A, et al. Vaccination with amyloid-β peptide induces autoimmune encephalomyelitis in C57/BL6 mice. *Brain* 2003;**126**:285−91.
6. D'Andrea MR, Nagele RG. Morphologically distinct types of amyloid plaques point the way to a better understanding of Alzheimer's disease pathogenesis. *Biotech Histochem* 2010;**85**(2):133−47.
7. D'Andrea MR. *Bursting neurons and fading memories*. New York: Elsevier Press; 2014.
8. https://www.youtube.com/watch?v = _NTaGjQow1c [accessed 07.10.15].
9. Parri R. Book review: bursting neurons and fading memories. An alternative hypothesis of the pathogenesis of Alzheimer's disease. *Biotech Histochem* 2015;**90**(6):495−6.
10. Laird FM, Cai H, Savonenko AV, Farah MH, He K, Melnikova T, et al. BACE1, a major determinant of selective vulnerability of the brain to amyloid-β amyloidogenesis, is essential for cognitive, emotional, and synaptic functions. *J Neurosci* 2005;**25**(50):11693−709.
11. http://www.alzheimers.org.uk/site/scripts/documents_info.php?documentID=1764 [accessed 18.09.15].
12. Lange-Asschenfeldt C, Kojda G. Alzheimer's disease, cerebrovascular dysfunction and the benefits of exercise: from vessels to neurons. *Exp Gerontol* 2008;**43**(6):499−504.
13. Zlokovic B. Review. The blood−brain barrier in health and chronic neurodegenerative disorders. *Neuron* 2008;**57**:178−201.
14. Amiel SA. The effects of *Bordetella pertussis* vaccine on cerebral vascular permeability. *Br J Exp Pathol* 1976;**57**:653−62.

15. de la Torre JC. Review. Alzheimer disease as a vascular disorder: nosological evidence. *Stroke* 2002;**33**(4):1152−62.
16. http://www.alzheimersweekly.com/2013/09/the-bad-good-in-statins-alzheimers.html [accessed 01.10.15].
17. Román GC. Review. Vascular dementia: distinguishing characteristics, treatment, and prevention. *JAGS* 2003;**51**:S296−304.
18. Euser SM, Hofman A, Westendorp RGJ, Breteler MMB. Serum uric acid and cognitive function and dementia. *Brain* 2009;**132**(2):377−82.
19. Feke GT, Hyman BT, Stern RA, Pasquale LA. Retinal blood flow in mild cognitive impairment and Alzheimer's disease. *Alzheimer's Dementia Diagnosis Assessm Dis Monitor* 2015;**1**:144−51.
20. Ujiie M, Dickstein DL, Carlow DA, Jefferies WA. Blood−brain barrier permeability precedes senile plaque formation in an Alzheimer disease model. *Microcirculation* 2003;**10**:463−70.
21. Zhang W, Huang W, Jing F. Review. Contribution of blood platelets to vascular pathology in Alzheimer's disease. *J Blood Med* 2013;**7**:141−7.
22. Peña VB, Bonini IC, Antollini SS, Kobayashi T, Barrantes FJ. Alpha 7-type acetylcholine receptor localization and its modulation by nicotine and cholesterol in vascular endothelial cells. *J Cell Biochem* 2011;**112**(11):3276−88.
23. Peng X, Gerzanich V, Anand R, Whiting PJ, Lindstrom J. Nicotine-induced increase in neuronal nicotinic receptors results from a decrease in the rate of receptor turnover. *Mol Pharmacol* 1994;**46**(3):523−30.
24. Wirths O, Multhaup G, Bayer TA. Review. A modified β-amyloid hypothesis: intraneuronal accumulation of the β-amyloid peptide—the first step of a fatal cascade. *J Neurochem* 2004;**91**:513−20.
25. Picciotto MR, Zoli M. Review. Nicotinic receptors in aging and dementia. *J Neurobiol* 2002;**53**:641−55.
26. Dineley KT, Bell KA, Bui D, Sweatt JD. Beta-amyloid peptide activates α7 nicotinic acetylcholine receptors expressed in *Xenopus* oocytes. *J Biol Chem* 2002;**277**(28):25056−61.
27. De Jonge WJ, Ulloa L. The α7 nicotinic acetylcholine receptor as a pharmacological target for inflammation. *Br J Pharmacol* 2007;**151**:915−29.
28. Zerbinatti CV, Wahrle SE, Kim H, Cam JA, Bales K, Paul SM, et al. Apolipoprotein E and low density lipoprotein receptor-related protein facilitate intraneuronal Aβ42 accumulation in amyloid model mice. *J Biol Chem* 2006;**281**(47):36180−6.
29. Dziewczapolski G, Glogowski CM, Masliah E, Heinemann SF. Deletion of the α7 nicotinic acetylcholine receptor gene improves cognitive deficits and synaptic pathology in a mouse model of Alzheimer's disease. *J Neurosci* 2009;**29**(27):8805−15.
30. Kanekiyo T, Liu C-C, Shinohara M, Li J, Bu G. LRP1 in brain vascular smooth muscle cells mediates local clearance of Alzheimer's amyloid-β. *J Neurosci* 2012;**32**(46):16458−65.
31. D'Andrea MR, Nagele RG. Targeting the alpha 7 nicotinic acetylcholine receptor to reduce amyloid accumulation in Alzheimer's disease pyramidal neurons. *Curr Pharm Des* 2006;**12**:677−84.
32. D'Andrea MR, Lee DHS, Wang H-Y, Nagele RG. Targeting intracellular Aβ42 for Alzheimer's disease drug discovery. *Drug Dev Res* 2002;**56**:194−200.
33. Perry G, Friedman R, Shaw G, Chauo V. Ubiquitin is detected in neurofibrillary tangles and senile plaque neurites of Alzheimer disease brains. *PNAS* 1987;**84**:3033−6.
34. Goedert M. Review. Oskar Fisher and the study of dementia. *Brain* 2009;**132**:1102−11.
35. D'Andrea MR, Nagele RG, Wang H-Y, Peterson PA, Lee DHS. Evidence that neurones accumulating amyloid can undergo lysis to form amyloid plaques in AD. *Histopathology* 2001;**38**:120−34.

36. Ginsberg SD, Crino PB, Hemby SE, Weingarten JA, Lee V, Eberwine JH, et al. Predominance of neuronal mRNAs in individual Alzheimer's disease senile plaques. *Ann Neurol* 1999;**45**(2):174−81.

37. D'Andrea MR, Nagele RG, Gumula NA, Reiser PA, Polkovitch DA, Hertzog BM, et al. Lipofuscin and Aβ42 exhibit distinct distribution patterns in normal and Alzheimer's disease brains. *Neurosci Lett* 2002;**323**:45−9.

38. Emmerling MR, Watson MD, Raby CA, Spiegel K. The role of complement in Alzheimer's disease pathology. *Biochim Biophys Acta* 2000;**1502**:158−71.

39. Mackenzie IRA, Munoz DG. Nonsteroidal anti-inflammatory drug use and Alzheimer-type pathology in aging. *Neurology* 1998;**50**(4):986−90.

40. Lue L-F, Yan SD, Stern DM, Walker DG. Review. Preventing activation of receptor for advanced glycation endproducts in Alzheimer's disease. *Curr Drug Targets—CNS Neurol Dis* 2005;**4**:249−66.

41. Cataldo AM, Nixon RA. Enzymatically active lysosomal proteases are associated with amyloid deposits in Alzheimer brain. *PNAS* 1990;**87**:3861−5.

42. Ohsawa K, Irino Y, Nakamura Y, Akazawa C, Inoue K, Kohsaka S. Involvement of P2X4 and P2Y12 receptors in ATP-induced microglial chemotaxis. *Glia* 2007;**55**(6):604−16.

43. Idzko M, Hammad H, van Nimwegen M, Kool M, Willart MAM, Muskens F, et al. Extracellular ATP triggers and maintains asthmatic airway inflammation by activating dendritic cells. *Nat Med* 2007;**13**:913−19.

44. Honda S, Sasaki Y, Ohsawa K, Imai Y, Nakamura Y, Inoue K, et al. Extracellular ATP or ADP induce chemotaxis of cultured microglia through Gi/o-Coupled P2Y receptors. *J Neurosci* 2001;**21**(6):1975−82.

45. Schraen-Maschke S, Sergeant N, Dhaenens C-M, Bombois S, Deramecourt V, Caillet-Boudin ML. Review. Tau as a biomarker of neurodegenerative diseases. *Biomarkers Med* 2008;**2**(4):363−84.

46. Giannakopoulos P, Herrmann FR, Bussiere T, Bouras C, Kovari E, et al. Tangle and neuron numbers, but not amyloid load, predict cognitive status in Alzheimer's disease. *Neurology* 2003;**60**(9):1495−500.

47. Arriagda PV, Growdon JH, Hedley-Whyte ET, Hyman BT. Neurofibrillary tangles but not senile plaques parallel duration and severity of Alzheimer's disease. *Neurology* 1992;**42**(3):631.

48. Stancu I-C, Vasconcelos B, Terwel D, Dewachter I. Models of β-amyloid induced Tau-pathology: the long and "folded" road to understand the mechanism. *Mol Neurodegener* 2014;**9**:51.

Glossary

Acetylcholine (Ach) A transmitter of the cholinergic neurons secreted at the ends of nerve fibers to propagate a signal and was one of the first neurotransmitters to be identified. In the cortex of the brain, Ach plays a role in attention, learning, and memory. Since the neurons that synthesize Ach die, less is present in the AD brain, and therefore, inhibiting the enzyme that breaks down Ach is a commonly used therapy to treat the symptoms of AD.

Adenosine diphosphate (ADP)/adenosine triphosphate (ATP) An intracellular molecule. Referred to as the molecular currency of the cells and used as a coenzyme to drive many of the intracellular metabolic processes in all living cells such as formation of nucleic acids, transmission of nerve impulses, and muscle contraction. Extracellular ATP serves as a danger signal to alert the immune system of tissue damage.

Alpha 7 nicotinic acetylcholine (α7) receptor A member of the neurotransmitter-gated ion channel superfamily. In neurons, α7 proteins assemble as a homopentamer composed of five individual subunits in presynaptic nerve terminals. This ligand-gated calcium receptor is stimulated by nicotine, acetylcholine, and Aβ, and is implicated in long-term memory. α7 has a very wide distribution beyond the nervous system to include inflammatory cells and cells of the vascular system.

Alzheimer's disease A progressive, neurodegenerative disorder of the brain and the most common underlying cause of dementia. Named after Dr Alois Alzheimer based on the autopsy of a subject with severe memory problems, confusion, and difficulty in understanding questions, neuropathological features include dense deposits of amyloid and neurofibrillary tangles in affected neurons. Pathology shows frequent Aβ amyloid deposition and neurofibrillary tangles.

Amyloid precursor protein intracellular domain (AICD) APP intracellular domain of 57- and 59-residue-long COOH-terminal fragments. This membrane product of APP is subsequently cleaved by γ-secretase to produce the Aβ peptide and another AICD fragment. Different AICD levels as a result of secretase activity may contribute to early pathophysiological mechanisms in AD. However, the relevance of AICD to AD is unclear.

Amyloid precursor protein-like (APPL) The APP is also conserved in invertebrates. *Drosophila* possesses a single APP homologue called APP like, or APPL, which is a neuronal-specific protein expressed in most, if not all, neurons throughout development and adult life. APPL is also a modulator required for axonal outgrowth.

Amyloid/APP/Aβ Broadly described as insoluble aggregates of protein. Of the various types of amyloid, the beta-amyloid (Aβ) form is associated with Alzheimer's disease that is a cleaved product from the amyloid precursor protein (APP). The Aβ is further cleaved to produce the Aβ42 form that is implicated in the neuropathology of Alzheimer's disease chiefly because of its propensity to aggregate (AβPP = APP).

Apolipoprotein (ApoB, ApoE, ApoE4) This protein combines with lipids in the body to form molecules called lipoproteins. ApoE is a major component of the very-low-density lipoprotein type and removes excess cholesterol from the blood to transport to the liver for processing, and therefore plays an important role in maintaining normal cholesterol levels. *APOE-ε4* allele was identified as one of the first genetic risks for AD. Of the forms of ApoE, having a single copy of the *APOE-ε4* allele increases the risk of AD, which increases considerably by having two alleles of *APOE-ε4*.

Apoptosis A regulated process of cell death that is in contrast to cells dying by necrosis, which happens from acute injury. In comparison to cell death by necrosis, apoptotic cells produce cell fragments called apoptotic bodies that phagocytes, like microglia, are able to engulf and quickly remove the contents of the cell before they cause further damage to the surrounding area.

Astrocytes Star-shaped glial cells of the brain and spinal cord. They support the endothelial cells that aid in the maintenance of the blood—brain barrier. Other functions include a role in repair and scarring in the brain, as observed in the areas of the dense-core plaques, and are also responsible for the maintenance of extracellular ion balance.

β-site APP-cleaving enzyme (BACE) βeta-secretase is a transmembrane aspartic protease. BACE2, a protease homologous to BACE1, was also identified, and together the two enzymes define a new family of transmembrane aspartic proteases. BACE1 exhibits all the functional properties of β-secretase, and as the key enzyme that initiates the formation of Aβ, BACE1 is an attractive drug target for AD.

Beilschowsky silver stain About a 100-year-old staining method that uses silver to nonselectively stain proteins in tissues and specific to the brain. Silver staining is used to demonstrate the presence of neurofibrillary tangles, nerve fibers, and senile plaques in Alzheimer brain tissues.

Blood—brain barrier (BBB) A physiological barrier comprising of tightly joined endothelial cells, astrocytes, pericytes, and smooth muscle cells to prevent certain substances in the bloodstream from entering the brain. The barrier may be the first pathological event and the root cause of Alzheimer's disease that leads to the unregulated entry of Aβ42 and neuronal-specific autoantibodies into the brain.

Blood—retina barrier (BRB) The BRB is composed of both an inner and an outer barrier. The outer BRB refers to the barrier formed at the retinal pigment epithelial cell layer and functions, in part, to regulate the movement of solutes and nutrients from the choroid to the sub-retinal space. In contrast, the inner BRB, similar to the BBB, is located in the inner retinal microvasculature and comprises the microvascular endothelium, which lines these vessels. The tight junctions located between these cells mediate highly selective diffusion of molecules from the blood to the retina and the barrier is essential in maintaining retinal homeostasis. The barrier becomes more porous in patients with diabetic retinopathy, which frequently occurs as the result of diabetes. It was proposed that the initial pathological events in the BBB mimic those in the BRB; therefore, high-resolution eye examinations could indirectly assess the BBB as additional AD risk factor.

Cathepsin D One of a family of cathepsin proteolytic enzymes that are located in the cell lysosomes and are chiefly involved in peptide synthesis and protein degradation.

Centromere DNA A region of DNA typically found in the middle of a chromosome that represents large blocks of repetitive DNA.

Cerebral amyloid angiopathy (CAA) CAA is a neurological condition in which amyloid builds up on or in vascular walls in the arteries of the brain. CAA is associated with a plethora of abnormalities and malfunctioning of the cerebrovasculature including ruptures of the vessel walls causing microhemorrhages and lesions of the BBB. These result in inflammation, vascular edema, and uncontrolled influx of peripheral blood components into the brain parenchyma.

Cerebral spinal fluid (CSF) CSF is a colorless fluid that bathes the brain and spine in nutrients and eliminates waste products. It also serves as a cushion to help prevent injury in the event of trauma. CSF is produced by the ependymal cells of the choroid plexuses of the ventricles in the brain.

Clathrin and Dynamin A protein that is essential for clathrin-mediated endocytosis. An adaptor protein that forms the primary component of the vesicle coating complex, while the dynamin, a GTPase, is responsible for the membrane cleavage to release the invaginated vesicle from the plasma membrane.

Congo red The staining method of choice to detect amyloid in tissue sections and was initially produced to stain textile fibers.

DAPI A blue fluorescent nuclear dye that binds strongly to the A-T rich sequences in DNA. DAPI is short for 4′-6′-diamidino-2-phenylindole.

Dementia A general term to describe a decline in mental ability (eg, thinking, memory, reasoning) severe enough to interfere with daily life; a progressive decline in cognition associated with an inability to perform normal activities owing to the cognitive deterioration.

Dense-core and diffuse amyloid plaques The former is also referred to as senile or neuritic plaques in the Alzheimer's disease brain. Their origin is a matter of debate. Most believe that extracellular deposition of Aβ from neurons as the diffuse plaque will aggregate over time with the help of inflammatory cells to mature into senile or neuritic plaques, which is based on the amyloid cascade hypothesis. Others strongly believe this is inaccurate, that these dense-core amyloid plaques are the result of a lysed neuron, which is based on the "Inside-Out" or Lytic Model, and that the diffuse plaques are simple areas of pooled vascular leakage of amyloid.

Endosome An acidic transitional compartment in equilibrium with the cell surface, trans-Golgi network, and lysosome, which compartmentalizes proteolytic function.

Entorhinal cortex An area in the brain that functions as the major input and output structure of the hippocampus and also functions in memory and navigation. The entorhinal cortex is one of the first two areas of the brain vulnerable to damage at the early stages of Alzheimer's disease.

Familial Alzheimer's disease (FAD) Also referred to as early-onset AD with patients developing symptoms in their 30s or 40s. Although an uncommon form of AD (∼2–3%), the genetics of FAD implicate mutations in the Aβ processing pathway enzymes that leave to AD, upon which the amyloid cascade hypothesis is based.

Formic acid A carboxylic acid used to pretreat formalin-fixed brain tissues before the immunohistochemical methods to help the antibodies penetrate the antigen or target.

Glial fibular acetic protein (GFAP) A specific intermediate filament found in astrocytes and used as a specific marker to identify astrocytes in immunohistochemical assays.

Gliosis An inflammatory response by the glial cells (eg, astrocytes, microglia) to damage areas in the brain. Initial response includes the migration of the microglia to the site of the injury, followed by the production of a dense fibrous network of astrocytic processes producing the glial scar to isolate and sequester the damage from the unaffected areas in the brain.

Hippocampus Part of the brain involved in memory forming, organization, and storing, spatial navigation, and important in forming new memories and associating them with emotions and senses. The hippocampus is one of the first two areas of the brain vulnerable to damage in the early stages of Alzheimer's disease.

Hypoperfusion Decreased blood flow through an organ and if prolonged, may result in permanent cellular dysfunction and cell death.

Immunohistochemistry (IHC) The process of detecting targets or antigens in cells or tissues through the use of specific antibodies and a detection system to color indirect presence of the target or antigen. IHC can detect the presence of 1, 2, or 3 simultaneous targets and is limited to a number of unique colors available and can be used in conjunction with other staining methods.

"Inside-Out" ("Bursting") hypothesis, now Lytic Model Proposes that the initial neuronal death in Alzheimer's disease is caused by unregulated intracellular accumulation of Aβ42 in neurons through the α7 receptor, leading to neuronal lysing and the development of dense-core plaques. This excess accumulation of Aβ42 in the brain is attributed to a dysfunctional blood−brain barrier, which should otherwise regulate and minimize the passage of amyloid. Once neuronal lysing and the formation of plaques begin, neuroinflammation and additional neuronal death follow.

In situ hybridization The process of detecting nucleic acid targets (eg, DNA, RNA) in cells or tissues using labeled complementary probes to produce a color at its location.

Integrins The integrins are transmembrane cell adhesion receptors. They regulate clathrin-mediated internalization of Aβ peptides at the neuronal synapses and are associated with lipid rafts, which are microdomains of signaling and trafficking areas in the plasma membrane.

Lipofuscin An intracellular yellowish-brown, nondegradable pigment that remains from the breakdown of cellular components that remains in cells that do not divide. In the "Inside-Out" hypothesis, this indigestible material remains at the epicenter or middle of the dense-core plaque as the neuron dies.

Low-density lipoprotein receptor-related protein (LRP) LRP-1 a highly conserved receptor that binds numerous types of ligands, including Aβ, and ApoE. LRP is a member of a family of structurally closely related transmembrane proteins, which participate in a wide range of physiological processes, including the regulation of lipid metabolism, protection against atherosclerosis, neurodevelopment, and transport of nutrients and vitamins.

Lysis The breaking down, dissolution or destruction of a cell by enzyme or other mechanisms to compromise the cells' integrity.

Lysosomes Membrane-bound cell organelles containing enzymes (as many as 50 types), which are able to break down all kinds of biomolecules including proteins, lipids, carbohydrates, nucleic acid, and cellular debris. The membrane protects the rest of the cells from the degradative enzymes. In the "Inside-Out" hypothesis, it is the uncontrolled release of these enzymes as the neurons die that created the spherically shaped dense-core amyloid plaques.

Microglia A type of glial cell in the brain that functions as the resident macrophage or inflammatory cell.

Microtubule-associated protein 2 (MAP2) Intracellular protein in neuronal dendrites that stabilizes microtubule growth by crosslinking microtubules with intermediate filaments.

Mild cognitive impairment (MCI) Recognized as an intermediate stage of mild impairment between the expected cognitive decline in normal aging and the more serious decline of dementia. May be perceived as a harbinger of Alzheimer's disease at the rate of 10–15% per year.

Mini-mental state examination (MMSE, or Folstein test) A 30-point questionnaire that is used extensively in clinical and research settings to measure cognitive impairment. With a maximum of 30 points, a score between 20 and 24 suggests mild dementia, 13–20 suggests moderate dementia, and less than 12 indicates severe dementia.

Multivesticular bodies (MVBs) Considered to be late endosomes formed from early endosomes by membrane invaginations that generate inner vesicles with lower pH ($\sim$5.5); may impair MVB sorting within the neuron by inhibiting the ubiquitin-proteasome system.

Necrosis An irreversible form of cell injury that is caused by a number of factors and results in the unregulated digestion of its cellular components, much like what has been described for the origin of the dense-core plaques from neurons that overingest Aβ42. Necrosis can also be active in the inflammatory system.

Neprilysin A zinc-dependent metalloprotease expressed in a myriad of tissues. In the neurons, neprilysin is reported to degrade Aβ, leading to a reduction of intraneuronal Aβ. Neprilysin is also reported to decline with age. Since neprilysin is localized to synapses, its age-related decline could explain the synaptic accumulation of Aβ42 that was observed in both AD transgenic mice and human AD brains.

Neuritic (senile) plaques Large extracellular aggregates of the amyloid Aβ peptide surrounded by dystrophic neurites (dendrites) containing aggregated tau. Also see dense-core amyloid plaques.

Neurofibrillary tangles (NTFs) Insoluble, intracellular aggregates of hyperphosphory-lated tau protein and cleaved forms of the microtubule-associated protein tau found within some of the neurons in the AD brain. NFTs are stained darkly with silver staining methods. Also see tau.

Neuronal-specific nuclear protein (NeuN) Expressed in most neuronal cell types throughout the nervous system and often used as an immunohistochemical marker to identify neurons.

Neuropil (neuropile) A general term to describe any area in the nervous system that includes the fibrous network of nerve fibers and glial cells between the neurons.

Neuregulin 1 An axonal signaling molecule critical for regulating myelination and therefore synaptic plasticity. Neuregulin 1 is also another substrate of BACE1.

N-methyl-D-aspartate (NMDA) NMDA receptor is a glutamate receptor and ion channel protein found in nerve cells and is critical in synaptic plastic and therefore learning and memory. NMDA receptors regulate clathrin-mediated internalization of Aβ peptides at the neuronal synapses and are associated with lipid rafts, which are microdomains of signaling and trafficking areas in the plasma membrane.

Nonsteroidal anti-inflammatory drugs (NSAIDs) NSAIDs are a category of medications that include the salicylate, propionic acid, acetic acid, fenamate, oxicam, and the cyclooxygenase (COX)-2 inhibitor classes. NSAIDs also reduce prostaglandins, which are a family of chemical mediators produced by cells to promote inflammation.

Parenchyma Composed of the neurons embedded in the framework of glial cells (microglia, astrocytes, and oligodendrocytes) and blood vessels.

Perikaryon The cell body of the neuron that houses the nucleus and many of its organelles, and an area reported to store Aβ42 over time.

Presenilins (PSEN, or PS) Presenilins are transmembrane proteins with eight transmembrane domains, located mainly in the endoplasmic reticulum (ER), Golgi, ER–Golgi intermediate structures, and synaptic terminals as detected by electron microscopy. Presenilins also mediate the processing of APP into its fragments. Mutations in the presenilins genes, presenilin 1 on human chromosome 14, and presenilin 2 on chromosome 1, in subjects with FAD also increase Aβ production and aggregation.

Proteolysis The enzymatic breakdown of proteins into smaller parts (amino acids, or polypeptides) through the hydrolysis of peptide bonds.

Pyknosis/pyknotic The irreversible condensation or shrinking of the nuclear chromatin representing a degenerating cell.

Receptor for advanced glycosylation endproducts (RAGE) Another receptor that binds to Aβ is RAGE, which is a multiligand cell-surface receptor expressed by neurons, and a variety of cells including the inflammatory cells of the brain (astrocytes and microglia) and vascular cells (endothelial, pericytes, and smooth muscle cells). RAGE binds monomeric, oligomeric, and even fibrillar Aβ at the surface of neurons, and then co-internalizes with Aβ.

Secretases Proteases designed to cleave transmembrane proteins to release bioactive forms or metabolize proteins prior to degradation. Specifically, secretases act of APP to cleave the protein into three fragments by sequential cleavage by β-secretase and then by γ-secretase to produce Aβ.

Soluble amyloid precursor protein (sAPP) In the non-amyloidogenic pathway, APP is first cleaved by the α-secretase producing the soluble APP-α (sAPP-α) peptide that is secreted into the extracellular medium.

Sortilin-related receptor 1 (SORL1) Neuronal SORL1 is an endocytic receptor implicated in the uptake of lipoproteins and proteases, and acts as a sorting receptor that can protect APP from trafficking to late endosomes and from processing into Aβ. Reduced function of SORL1 led to increased sorting of APP into Aβ-generating compartments.

Synaptic hypothesis A variation of the initial deposition model pinpoints the area of the neuronal-secreted amyloid to the synapse suggesting the extracellular release of Aβ from degenerating neurites might upregulate Aβ42 within adjacent synaptic compartments, leading to the synaptic spread of AD.

Synaptic plasticity (long-term depression (LTD) and long-term potentiation (LTP)) The process by which neuronal synapses modulate their strength and form new connections with other neurons that play particularly important roles in learning and memory as well as response to injury and disease. The strengthening of synaptic connections also referred to as LTP, while LTD is an activity-dependent reduction or weakening of synaptic connections.

Tau Tau protein is mostly present in neurons as compared to nonneuronal cells. It is a microtubule-associated protein that stabilizes the axonal microtubules. When tau is abnormally phosphorylated, it can form neurofibrillary tangles, which are one of the neuropathological features in the Alzheimer's disease brain.

Ubiquitin An intracellular protein found in almost all eukaryotic cells (ubiquitously) that combines with obsolete proteins to make them susceptible to degradation. Also involved in the cellular processes of cell-cycle regulation, DNA repair, cell growth, and others.

VaD Defined as loss of cognitive function resulting from ischemic, hypoperfusive, or hemorrhagic brain lesions due to cerebrovascular disease or cardiovascular pathology.

Western analysis Also known as western blotting, western analysis is a technique used to identify proteins in a sample lysate of digested and grinded up cells or tissues, with the use of antibodies, and gel electrophoresis.